Agricultural biodiversity

This book presents part of the findings of the international project "People, Land Management, and Environmental Change", which was initiated in 1992 by United Nations University. From 1998 to 2002 the project was supported by Global Environment Facility with the United Nations Environment Programme as implementing agency and the United Nations University as executing agency.

The views expressed in this book are entirely those of the respective authors, and do not necessarily reflect the views of the Global Environment Facility, the United Nations Environment Programme, and the United Nations University.

Agricultural biodiversity in smallholder farms of East Africa

Edited by Fidelis Kaihura and Michael Stocking

United Nations
University Press

TOKYO · NEW YORK · PARIS

UNEP

GEF

United Nations University Press
The United Nations University, 53-70, Jingumae 5-chome,
Shibuya-ku, Tokyo, 150-8925, Japan
Tel: +81-3-3499-2811   Fax: +81-3-3406-7345
E-mail: sales@hq.unu.edu (general enquiries): press@hq.unu.edu
www.unu.edu

United Nations University Office in North America
2 United Nations Plaza, Room DC2-2062, New York, NY 10017, USA
Tel: +1-212-963-6387   Fax: +1-212-371-9454
E-mail: unuona@ony.unu.edu

United Nations University Press is the publishing division of the United Nations University.

Cover design by Rebecca S. Neimark, Twenty-Six Letters

Printed in Hong Kong

UNUP-1088
ISBN 92-808-1088-X

Library of Congress Cataloging-in-Publication Data

Agricultural biodiversity in smallholder farms of East Africa / edited by Fidelis Kaihura and Michael Stocking.
    p.   cm.
Includes bibliographical references and index.
ISBN 92-8081088-X
1. Agrobiodiversity—Africa, East. 2. Agrobiodiversity conservation—Africa, East. 3. Farms, Small—Africa, East. I. Kaihura, Fidelis. II. Stocking, Michael.
S494.5.A43   A45   2003
333.95'16'09676—dc22   2003017881

Contents

Part I: Introduction to agricultural biodiversity in East Africa

v

Part II: Components of agricultural biodiversity

Part III: Farmers' perspectives

Part IV: Policy recommendations

List of tables and figures

List of acronyms

ABS	access and benefit sharing
AEZ	agricultural ecological zone
AU	African Union
BAG	biodiversity action group
BUDEG	Bushwere Development Group
BUNUHOGAFA	Bushwere Nursery and Home Garden Farmers' Association
BUZECIA	Bushwere Zero-grazing and Crop Integration Association
CBD	Convention on Biological Diversity
CBO	community-based organization
CCA	canonical correspondence analysis
DEO	district environment officer (Uganda)
EAPLEC	East Africa PLEC
FPR	farmer participatory research
FSR	farming systems research
FYM	farmyard manure
GDP	gross domestic product
GEF	Global Environment Facility
GR	genetic resources
IPR	intellectual property rights
ITK	indigenous technical knowledge
IUCN	International Union for the Conservation of Nature
KARI	Kenya Agricultural Research Institute
LUT	land-use type
MAAIF	Ministry of Agriculture, Animal Industry, and Fisheries (Uganda)

MPEFA	Mwizi PLEC Experimenting Farmers' Association
MWLE	Ministry of Water, Lands, and Environment (Uganda)
NAADS	National Agricultural Advisory Services (Uganda)
NEMA	National Environment Management Authority (Uganda)
NGO	non-governmental organization
NUWA	National Urban Water Authority (Tanzania)
OP	operational programme (CBD)
PEA	participatory extension approach
PEAP	Poverty Eradication Action Plan (Uganda)
PLA	participatory learning and action
PLEC	People, Land Management, and Environmental Change project
PMA	plan for the modernization of agriculture (Uganda)
PNA	participatory needs assessment
PRA	participatory rural appraisal
PSA	participatory situation analysis
PTD	participatory technology development
PTDD	participatory technology development and dissemination
RELMA	Bushwere Regional Land Management project
SFMP	soil fertility management practices
SWC	soil and water conservation
SWOT	strengths, weaknesses, opportunities, and threats
TANESCO	Tanzania Electrical Supplies Company
UNDP	United Nations Development Programme
UNEP	United Nations Environment Programme
UNU	United Nations University
UPDF	Uganda People's Defence Force

Foreword

Biodiversity is a key global concern of the international community. Extinction rates of species and rare varieties of biota are reportedly accelerating to alarming levels. Especially in tropical developing countries where biodiversity is naturally rich but human beings are poor, conservation of biodiversity cannot be confined to protected areas. Conservation must address landscapes – or the 94 per cent of the earth's surface that is not designated as a protected area – that are managed for plant cultivation, pastoralism, forests, recreation, and other occupations that support human livelihoods. That is why this book is to be applauded for bringing to our attention the diversity of ways that biodiversity is being protected on smallholder farms in a part of Africa long famed for the variety and richness of its flora and fauna. East African smallholder farmers are pre-eminently the guardians of both the biodiversity surrounding them and the knowledge to manage it.

The United Nations University has been proud to initiate one of the foremost projects on agricultural biodiversity, called People, Land Management, and Environmental Change (PLEC). Originating from one of the UNU's thematic areas as the UN "think-tank" on environment and sustainable development, PLEC has developed a global network of 27 demonstration sites of good practice in protecting biodiversity that is beneficial to society and provides food, fuel, and shelter, as well as protecting biota that could be beneficial to future generations. However, PLEC is not just about protecting biodiversity; it is also about supporting human livelihoods. At the core of this objective is the retention of the knowledge of local communities on how to manage their biodiversity in ways beneficial both to the local people themselves and to the global community. Much of this local knowledge on how to manage plants and complex biophysical environments is fast disappearing. PLEC and the UNU, with funding support from the Global Environment Facility and the United Nations Environment Programme, are making a small but vital contribution to our global understanding of how to manage biodiversity.

The UNU would like to acknowledge the contribution of farmers worldwide to biodiversity conservation, and the interest of developing country scientists (now well over 200 in the PLEC project) to work with local experts to the common good. This book of experiences from East Africa will provide a rich source of knowledge for all scholars on positive ways forward to achieve the goal of an environmentally sustainable and protected planet earth.

Professor Motoyuki Suzuki
Vice-Rector
United Nations University

Tokyo, October 2002

Part I

Introduction to agricultural biodiversity in East Africa

1

Agricultural biodiversity in East Africa: Introduction and acknowledgements

Michael Stocking, Fidelis Kaihura, and Luohui Liang

Introduction

Throughout eastern Africa, broadly taken as extending from the Horn of Africa down through the former British colonies of East-Central Africa to Mozambique, there is a large diversity of farming systems, human societies, and ways of managing complex external pressures on sustainable land use. In Ethiopia, for example, Konde *et al.* (2001) document how in the highly populated parts of Wolataya, farmers have created intensive gardens, based largely on intricate soil fertility management practices. Biographies of change in attitudes and activities of farmers reveal complex forcing mechanisms that have resulted in today's pattern of land use, with many important implications for the development of policy and future practice.

This example from the northern part of eastern Africa is but one of many accounts of the complex interactions between land users and their broader environments from the whole region. This book focuses attention on the central part of eastern Africa. It is primarily about the investigations and experiences of colleagues and farmers in three countries – Kenya, Tanzania, and Uganda. These countries were chosen not because they have the best examples, but because they were clustered together by one project, People, Land Management, and Environmental Change (PLEC), in a so-called "hot-spot" of biodiversity, where the international community as represented by the Global Environment Facility felt that

additional attention was needed. The researchers have sought an understanding of the role of biodiversity on the agricultural lands of smallholder farmers. This opening chapter sets the scene by first providing an
overview of what is meant by "agricultural biodiversity" (or what this
project has termed "agrodiversity"). Then the chapter provides some
definitions of terms, which many find confusing. Thirdly it describes the
global PLEC project, with its East African cluster, that provided the
funding as well as the methodology for the work. There is then an overview of the book. Finally, acknowledgement is made of the many people
who contributed to the research, without whom this book could not have
been written. Fidelis Kaihura wrote this last section of this chapter in his
role as the nominated new cluster leader of PLEC for East Africa.

Agricultural biodiversity

Agricultural biodiversity is a topic that has only within the last decade
come to the fore as an issue worthy of special attention, study, and research. It describes the situation of biological diversity in areas of agricultural activity and land use. Since land use – or perhaps more exactly,
land abuse – is considered by most observers to be the major threat to
biological diversity, it may appear to be something of an enigma that
agricultural biodiversity should exist at all. If agricultural activities deplete biodiversity, then surely efforts to protect biodiversity should focus
on non-agricultural areas – forest reserves, wildlife sanctuaries, national
parks, and wilderness areas?

Yet because biodiversity is a global concern and because most productive areas of the world, which contain most of the globe's biodiversity,
are in areas of land use, agricultural biodiversity is far from being the
contradiction that narrow ecologists would see it to be. Indeed, on the
grounds that most of the "hot-spots" of biodiversity are intensively used
and support large and growing populations, an alternative argument
should prevail. It could be argued that agricultural biodiversity (or agrobiodiversity) is far more important than, say, conservation of protected
areas or of remnants of natural habitats in areas of land use (Wood and
Lenné 1999). Equally, there is probably more intrinsic biological diversity in areas of land use than in all the protected areas put together –
a claim that is probably impossible to verify, but nevertheless useful to
support the importance of agricultural biodiversity. Ecological and conservation purists might counter that argument by saying that this is un-
natural biodiversity, full of alien and invasive species. Land use has
destroyed the natural habitats, created biological seas of uniformity, and
even eradicated small niches of interest such as hedgerows and field

boundaries. They would have a point if one concentrated only on areas of commercial farming and forestry, where monocrops and single varieties prevail and single-species stands of trees line up in rows. However, the agricultural biodiversity dealt with in this book is under the guardianship of smallholder farmers, with diverse practices, interests, skills, and needs. They manage landscapes rich in species and intricate in their organization. Furthermore, they do not just protect many indigenous species, they conserve and manage plants and animals important to human beings. Yes, it is not the natural biodiversity of species and varieties, many of which have not yet been discovered and named. The authors of this book would claim it is a far more important and immediate biodiversity that also consists of indigenous skills, knowledge, and management. It involves plants and animals with use and non-use values, such as medicinal plants, local food crops, ornamental trees, and domestic animals.

To illustrate, one of the PLEC farmers in Arumeru believes in *matatu*, or growing three types of plant together. There seems to be no particular scientific rationale for these threesomes – he has many of them – but they work for him. He chooses the species carefully and he has a management strategy. It is important to document both the biota being managed and the way that land users have learnt how to manage it if we are to provide for food security, sustainable livelihoods, and human development and well-being. Sometimes we may have secretly believed what we are told to be myth. Sometimes scientists may add their formal knowledge, unavailable to local people, to create new approaches and new technologies. However, the whole assemblage of what is here termed "agricultural biodiversity" or "agrodiversity" is vital to be documented, examined, and understood. This book does this for demonstration sites set up under the PLEC project in three countries of East Africa: Kenya, Tanzania, and Uganda. Before outlining the structure of the book and describing the project itself, it is important briefly to define the key terms used.

A question of definition

The headline term used in the title of this book and this chapter is "agricultural biodiversity". The authors wanted to use "agrodiversity", because this means all aspects of biological diversity in areas of agricultural land use, plus the diverse ways that farmers manage the biota. However, "agrodiversity" is still not widely in use. In October 2002 the internet search engine Google came up with 561 references to "agrodiversity", many of which emanate from the PLEC project, and 274,000 for "agricultural biodiversity": nearly 500 times greater recognition for the second term than the first. The publisher of one of the previous books from the project (Brookfield *et al.* 2002) persuaded the authors at the last moment to change part of the title of the book from "agrodiversity" to "agricultural diversity", on the grounds that the preferred term lacked resonance with most people. The authors acquiesced then, and do so again now. However, the question of definitions needs to be made explicit.

Agrodiversity

Agrodiversity (also termed "agricultural diversity") is "the many ways in which farmers use the natural diversity of the environment for production, including not only their choice of crops but also their management of land, water and biota as a whole" (Brookfield and Padoch 1994: 9). The PLEC project has seen agrodiversity as essentially to mean "management diversity". This is related to "agricultural biodiversity" but encompasses much more. It includes the management of fields and soil fertility, as well as farms and landscapes. It takes in the application of agricultural technologies, crop rotations, soil and water conservation techniques, and weed and pest management. As Brookfield (2001) employs the term, agrodiversity also includes adaptation to resource degradation, and the employment of indigenous, adapted, and introduced knowledge to farming. Conceptually, "agrodiversity" is the broadest of the terms used to capture biological diversity and the diversity of management and organization at a variety of temporal and spatial scales.

Agricultural biodiversity

Also written in shorthand form as "agrobiodiversity", agricultural biodiversity means the diversity of useful plants in managed ecosystems. It has been defined as "the variety and variability of plants, animals and micro-organisms at genetic, species and ecosystem level" (Cromwell 1999: 11). Definitions of the term usually include the aspect of managing agricultural biodiversity and the importance of human intervention in the creation of an agriculturally biodiverse assemblage.

The global PLEC project and its East African cluster

Most biodiversity worldwide is managed by farmers and communities. While a large amount of crop genetic diversity is now collected and preserved in *ex situ* gene banks, farmers continue to conserve planting materials *in situ* in response to changing natural and social conditions.

This management of biodiversity in agricultural landscapes is receiving growing recognition and attention. However, the understanding of what farmers and communities can do to maintain and enhance biodiversity even in intensively cultivated areas is limited. Most funding for biodiversity conservation is used to support protected areas. There is now a strong demand at local, national, and international levels for participatory models of biodiversity management in agricultural ecosystems that embrace biodiversity for farmers' livelihoods.

The adoption of a work programme on agricultural biodiversity by both the Conference of the Parties to the Convention on Biological Diversity (CBD) and the Global Environment Facility (GEF) in 2000 marked a watershed in promotion of managing biodiversity in agricultural ecosystems. Decision V/5, adopted by the Conference of the Parties to the CBD at its fifth meeting in May 2000 in Nairobi, recommends efforts to "identify management practices, technologies and policies that promote the positive and mitigate the negative impacts of agriculture on biodiversity, and enhance productivity and the capacity to sustain livelihoods, by expanding knowledge, understanding and awareness of the multiple goods and services provided by the different levels and functions of agricultural biodiversity".

In advance of the CBD's recommendation, the United Nations University project on People, Land Management, and Environmental Change (PLEC) has spearheaded work on agricultural biodiversity. With support from the United Nations Environment Programme and the Global Environment Facility, it has brought together a large number of researchers and smallholder farmers for the identification, evaluation, and promotion of resource management systems that conserve biodiversity. At the same time, the protection should generate income and assist in coping with changes in social and natural conditions. The PLEC project operates through a global network of clusters that have been established in Ghana, Guinée, Kenya, Tanzania, Uganda, China, Thailand, Papua New Guinea, Brazil, Peru, Mexico, and Jamaica. Demonstration sites are located in priority agro-ecosystems in the margins of forests, semi-arid regions, mountains, and wetlands of globally significant biodiversity. Information about PLEC is available at www.unu.edu/env/plec/.

The PLEC concept

For thousands of years farmers have constantly modified their use and cultivation of biodiversity for food and livelihoods through learning, experiment, and innovation. Over this long history they have nurtured and managed a diversity of plants and animals, either wild or domesticated, and developed agrodiversity to harness various plants and animals. Equally, over this long history the types of agricultural land use have diverged. Especially with pressures emerging between and after the world wars to produce large quantities of cheap food, large-scale commercialization and mechanization of agriculture has become dominantly manifest. High-yielding varieties have replaced the huge diversity of local varieties and genotypes. The pressure to produce food has been inexorable in both developed and developing countries. However, for several reasons, pockets of small-scale agriculture and land use have remained. Often this is the more appropriate land use to feed high densities of rural populations in intensive home gardens, represented in East Africa classically by the Chagga home gardens on the slopes of Mount Kilimanjaro. These gardens are both productive and diverse, but except in certain very specific commodities, such as coffee, they do not contribute to the wider market. Elsewhere, physical isolation, as in mountain communities, has buffered land use from the pressures of the market. PLEC concentrates on these pockets of mainly small-scale, intensive, and diverse agricultural systems on the premise that they are worth conserving. It is not a question of whether agricultural development should go wholly towards high-yielding uniform systems of land use or to small-scale, low-input, diverse systems. Both are important, but the latter are more under threat. They contain value in their biodiversity and wealth of knowledge – aspects which perhaps have less resonance than food security and large grain storages, but which potentially have implications for the sustainability of both large-scale commercial agriculture and small-scale diverse agriculture. The challenge that PLEC addresses is how to conserve the diversity of techniques, species, varieties, and ways of organizing land use in complex land-use systems. In other words, how can agrodiversity be conserved?

What PLEC has found is that in the current trend towards uniformity in agricultural landscapes, a significant proportion of farmers and communities continue to develop agrodiversity – a dynamic patchwork of various land-use stages (such as annual cropping, orchard, agroforest, fallow, home garden, and boundary hedges). At the smallest scale, these land-use stages are specified as field types of land management that farmers recognize on the ground. The field types may be sequential management (such as seasonal variations of crops or varieties and shifting cultivation) and concurrent management (such as mixed cropping and agroforestry). Land-use stages are not fixed, but in a constant state of dynamic flux. At a PLEC site in Yunnan, China, for example, some farmers are expanding home gardens on to former rice paddy terraces for marketable vegetables, medicinal plants, and fruits. Other farmers are converting maize fields into an agroforestry association of native tree crops. The dynamic patchwork of transitional land-use stages and field

Table 1.1 Four elements of agrodiversity

Agrodiversity categories	Description
Biophysical diversity	The diversity of the natural environment including the intrinsic quality of the natural resource base that is used for production. It includes the natural resilience of the biophysical environment, soil characteristics, plant life, and other biota. It takes in physical and chemical aspects of the soil, hydrology, climate, and the variability and variation in all these elements.
Management diversity	All methods of managing the land, water, and biota for crop and livestock production, and the maintenance of soil fertility and structure. Included are biological, chemical, and physical methods of management.
Agrobiodiversity	All species and varieties used by or useful to people, with a particular emphasis on crop, plant, and animal combinations. It may include biota that are indirectly useful, and emphasizes the manner in which they are used to sustain or increase production, reduce risk, and enhance conservation.
Organizational diversity	This is the diversity in the manner in which farms are operated, owned, and managed, and the use of resource endowments from different sources. Explanatory elements include labour, household size, capital assets, reliance on off-farm employment, and so on.

Source: Adapted from Stocking (2002)

types mimics various stages of vegetation succession, maintains diversity of habitats, and harnesses biodiversity in space and time.

Agrodiversity emphasizes farmers' resource management of the whole landscape. It covers four elements: biophysical diversity, management diversity, agrobiodiversity, and organizational diversity (see Table 1.1), and their interactions. Farmers select and manage crops, but also choose, modify, and create the suitable microenvironments and soils for production. For example, natural or artificial forest is managed and conserved for raising snails, butterflies, or medicinal plants. Both domesticated and wild species are used for livelihoods. Farmers protect wild species through selective weeding and transplantation of wild tree seedlings.

Agrodiversity contains considerable potential for conservation of biodiversity, protection of important land-use systems, and control of land degradation as well as enhancement of food security and rural livelihoods. The role of agrodiversity in conserving biodiversity is demonstrated through a patchwork of various cropping systems, agroforestry systems, and forest systems that use and harness crop diversity. Evidence is accumulating that not only is there a wealth of good practice in many previously overlooked local systems for biodiversity conservation, but also such systems reduce land degradation risks and support local livelihoods. Agrodiversity practices provide nutrition and safe food, reduce production risk, and enhance the ability to cope with changes and mitigate disasters. Even under pressure for uniform production, many small farmers worldwide continue to practise agrodiversity for viable livelihoods. PLEC deliberately dwells on those "sustainable adaptations by small farmers to varied environments under growing population pressure and all other forms of stress ... principally through the high degree of structural, spatial and trophic, as well as species diversity that is involved" (Brookfield 1995: 389).

PLEC methodology

As agrodiversity is complex, it has taken time for PLEC to develop an effective methodology based on local expertise for identification and promotion of the agrodiversity in small farmers' agricultural systems (Liang 2002). Details of PLEC principles, working guidelines, and case studies on agrodiversity are described in Brookfield et al. (2002), Coffey (2000), and Stocking and Murnaghan (2000). The key components of the methodology are detailed below.

Demonstration sites

While the PLEC theme centres on agrodiversity, the approach to the theme began to change from research-oriented work to demonstration and capacity building in some of the clusters of the PLEC network in 1996, and subsequently in the whole network (Liang et al. 2001). Site selection is mainly based on regional biodiversity importance; threats to biodiversity by rapid change and land-use pressures; critical ecosystems based on national priorities and potentials; known examples of agrodiversity; and existing partnership with communities and availability of historical information. Some sites were those in which project members had previously worked, or were still working in connection with other projects.

However, setting up viable demonstration sites was difficult because the PLEC approach to "demonstration sites" was quite different. These sites are not so much physical places but rather people-centred processes, and coalition and partnership between scientists, farmers, local communities, and other stakeholders searching for sustainability on the ground. Some clusters before early 1999 had carried out essentially reconnaissance work along large transects extended over many kilometres and several agro-ecological zones. Some were overwhelmed by their own research agendas, and were unable to create genuine coalition with farmers and other local stakeholders. Those clusters were quickly advised to concentrate their work in more narrowly defined areas, and to benefit from the experiences of other successful clusters.

Participating clusters are selected on regional biodiversity importance. Most PLEC clusters overlap with "biodiversity hot-spots" in South-East Asia, East and West Africa, Central America, and South America. For viable demonstration site work, most clusters started with a few large areas of national priority and later narrowed down to a few defined sites. PLEC-Tanzania, for example, narrowed demonstration site areas to two landscape units, which were selected from an initial five units they had identified on both the windward and leeward slopes of Mount Meru in Arumeru district, Tanzania. To date, a total of 21 demonstration sites in eight GEF-supported countries (Brazil, China, Papua New Guinea, Uganda, Kenya, Tanzania, Ghana, and Guinée), and six sites in four UNU-supported countries (Peru, Mexico, Jamaica, and Thailand) are now established and operational. Further demonstration sites are in development, some of which are in response to popular demand from nearby communities.

Agrodiversity assessment

Since agricultural practices and their products vary in time and space and between households, PLEC assessment of biodiversity and its management in these production landscapes is based on land-use stages in the individual farms of sampled households. Traditional surveys of land use cannot catch the full picture of relay or rotational cropping and land-use management often practised by small farmers to maintain soil fertility and suppress pests. For example, farmers at the PLEC Tumam demonstration site, East Sepik province, Papua New Guinea, divide land management into two main periods. The first is the garden stage while the second is the fallow stage. The garden stage is further divided into three substages according to months after the initial clearing: *wah* (seven months), *yekene* (20 months), and *nerakase* (33 months). The fallow period consists of four substages according to years after the garden stage: *nerakase* (up to five years), *banande* (10 to 15 years), *loumbure* (15 to 20 years), and *loutinginde* (20 to 50 years). As the management passes from stage to stage, the species richness and species composition change. Land-use stages and their management as well as crop diversity are in constant flux in this dynamic mosaic. Temporal diversity is as important as spatial diversity.

The household is the basic unit of small-scale farming, though there is much cooperation between households, relatives, and the wider community. Variations between households in labour, resource endowments, and other conditions give rise to different approaches to managing their resources even within the same community. Some "expert farmers" manage resources much better than others. A general community-level survey would have failed to detect the difference in biodiversity and its management between households, as well as failing to spot expert farmers and their exceptional practices of agrodiversity.

As a result, the basic principle of the PLEC agrodiversity assessment is to stratify the landscape at demonstration sites. This is achieved through identifying land-use stages and field types so as to detect differences between land-use stages, field types, and households, especially when finding biodiversity-friendly and economically profitable systems for demonstration. In most cases clusters began transect surveys for identification of land-use stages and field types across the community and landscape, and sampled representative households and field types in their landholdings. Once a sample of representative households and field types in the community was selected, researchers with farmers conducted detailed inventories of plant species and household economy, monitored changes in plant species and household economy, and entered the information into a database for analysis.

Each component cluster of the PLEC network adapted the PLEC guidelines to their local situations. PLEC-Thailand, for example, employed different methods when the research moved from the reconnaissance stage to the detailed investigation stage. Details are provided in Table 1.2.

Promotion of agrodiversity

Since agrodiverse practices are well integrated with local ecosystems and livelihoods, they are site and household specific and cannot simply be copied to other environments, households, or communities. Promotion of these agrodiverse practices cannot be done through the conventional extension model of "transfer of technology". Moreover, farmers often obtain new ideas and technologies from exchanges with other farmers and prefer to see concrete results on the farm. As a result, PLEC promotes successful agrodiverse practices through on-farm demonstration and "farmer learning from expert farmer". In a typical on-farm demonstration, an expert farmer is facilitated to teach his or her practices to other farmers on his or her farm. Participating farmers are welcome to comment on the expert farmer's practices, and assimilate, change, or adapt those practices for their specific farms. The practices taught and formality used in a farmer-to-farmer demonstration depends on the choice of expert farmers as instructors. The formality may range from informal occasions to formal meetings. The informal occasions include family gatherings and labour exchange. The organization of demonstra-

Table 1.2 Field methods, tools, and approach with respect to expected outcomes of field activities of PLEC-Thailand

Field methods/ tools and approaches	Outcomes of field activities				
	Village landscape	Farming systems	Production systems	Household	Fields/plots
Mapping and PRA	Identification of land use and patterns of natural resources. Defining major production systems and identifying biophysical and organizational components of agrodiversity.				
Guideline "field type"		Characterizing the existing (distinct) farming and forest management practices with diverse crops and cropping systems. Grouping of common management practices.			
Agrodiversity checklist		Identifying sample plots as representative of the field type for direct observation and measurement.			
Household survey/field interview				Selection of sample households based on field types and potential for future demonstration. Collecting information on household socio-economic status and management of agrodiversity.	

Source: Adapted from Rerkasem *et al.* (2002)

tion activities can be facilitated directly by field researchers or through farmers' associations.

PLEC in East Africa

This subsection is based upon Kaihura *et al.* (1999). East Africa is renowned for its high natural biodiversity. From the forested mountains of western Uganda with its remnant populations of gorillas to the lush highlands of central Kenya with intensive agricultural systems and the remarkable endemism of Tanzania's equatorial mountain chain, the whole region has a wealth of flora and fauna as well as a rich natural biophysical diversity. Sharp contrasts over short distances in altitude, climate, vegetation, soils, and hydrology contribute to this diversity. Rainfall variability and soil fertility change markedly between villages, as shown in Embu in semi-arid Kenya (Tengberg *et al.* 1998), which has implications for the practices that local people follow, such as terracing or trash lines.

Diversity is also depicted in society and demography, with widely different ethnic groups such as the Masai, Kikuyus, Arumeru, and Mwizi peoples. Population growth is rapid, and densities vary from over 2,000 per km^2 to less than 20 in drier parts. Consequently, East Africa is a natural candidate for the study of how local agricultural and land-use systems interact with this natural biodiversity and how, in turn, the biodiversity contributes to local livelihoods. This mutual support between land use and livelihoods on one side and biological diversity on the other is a particular feature of East Africa. In the face of considerable external pressures, such as declining areas of land per person and rapidly changing market economies, land users are coping by exploiting biodiversity while at the same time demonstrating their protection of it if the circumstances are right.

East Africa PLEC's original objective was to examine the interaction between increasing population pressures, the intensified use of land, and associated effects such as migrations and rapid urbanization, and the various aspects of agrodiversity. East Africa is famous for the Machakos (Kenya) study entitled *More People, Less Erosion* (Tiffen, Mortimore, and Gichuki 1994), in which it was argued that intensification leads to more sustainable land-use practices and improved livelihoods. EAPLEC is now working on demonstration sites (see Figure 1.1), using the PLEC framework, from which it will be possible to gain detailed insights into farmers' strategies of managing biodiversity. The PLEC goal is to help farmers develop and conserve productive, sustainable, and biodiverse land-management systems. In East Africa these systems consist of a wide range of managed land uses from forests to agroforestry, dryland cropping to intensive vegetable production, and stall-fed livestock to rangeland.

Farmers' perspectives

Working closely with farmers, learning from them to find entry points for improvements on existing resource management models, and developing sustainable management techniques that also conserve biodiversity are all central to PLEC's work.

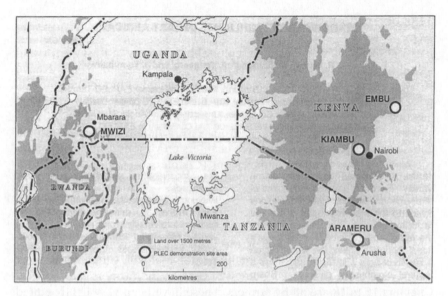

Figure 1.1 Location map of PLEC demonstration sites in East Africa

While approaches may vary between countries, farmer empowerment is always an ultimate objective in order to achieve lasting change and improvement in the management of the natural resource base. Activities towards this objective are necessarily diverse. They must be based on differences in land-use systems, management constraints, socio-cultural situations, and other related elements of agrodiversity. For Tanzania, Kenya, and Uganda, PLEC's sites were in a range of agro-ecologies, from the semi-arid to sub-humid and on to humid. While this set a natural bio-physical diversity and contrast in approaches to natural resources management, there were many other aspects that were much more similar. Demographic, cultural, economic, and social factors are all surprisingly similar between the three countries, but again with differences which are brought out in Part III of the book through the eyes largely of farmers.

In all cases, farmers were key participants at workshops and meetings and their input was encouraged and valued. For example, at the last East Africa PLEC workshop in 2001, farmer representatives from the three countries shared their experiences as PLEC farmers with other workshop participants. Part III of the book is therefore a collection of the perspectives of those farmers on PLEC's work. The overarching message is that, while there may be many similarities in driving forces and factors, there is no one blueprint approach to conservation and sustainable land management.

This book

Part I of this book introduces agricultural biodiversity in East Africa, and includes typical overview accounts from Uganda (Chapter 3) and Kenya (Chapter 4). Chapter 2 introduces the management of biodiversity and its position in current agendas. Following this introductory Part I, the book has three substantive sections. Part II looks at the components of agricultural biodiversity. These vary from a very detailed study of rainfall characteristics (Chapter 5) to three chapters on various aspects of botanical knowledge and plant management (Chapters 6 to 8), and a chapter on the role of livestock (Chapter 9). Also relevant to the components of agricultural biodiversity are Chapters 11 (socio-economic factors), 12 (production factors – in this case of bananas in Uganda), and 13 (land-use change).

A key feature of the book is the inclusion of a number of chapters either inspired by farmers or actually written by them. Chapter 14 covers the development of a methodology to capture the farmers' perspective. Chapters 15 to 18 are all by farmers. These have been very lightly edited to bring clarity, but they are kept largely in their original form in order to demonstrate how farmers feel about the subject and their interaction with researchers. Chapter 19 concludes Part III by reporting on farmer-led evaluations of soil management practices from Uganda.

The final Part IV of this book is on policy recommendations. The teams in Tanzania (Chapter 20) and Uganda (Chapter 21) devoted considerable efforts towards including policy-makers, inviting them whenever possible to visit the demonstration sites and participate with local people on understanding local needs. The results of these interactions between farmers, scientists, and politicians throw a fascinating insight on to how the work on agrodiversity may move forward to bringing real livelihood and food-security benefits to local people.

Acknowledgements

This book is dedicated to smallholder farmers of the PLEC demonstration sites of Embu in Kenya, Bushwere in Uganda, and Arumeru in Tanzania. The information in this book is a result of farmers' cooperation and the contribution of their invaluable time, knowledge, and experience in training other farmers, researchers, and extension staff, including policy-makers and other stakeholders, in diverse ways of managing agricultural biodiversity at farm and landscape level. They have also demonstrated the value of agricultural biodiversity in contributing to food security and rural livelihoods. Without this principal stakeholder, the farmers of East Africa, PLEC could not have operated.

The book is based on the proceedings of the East Africa Annual General Meeting in Arusha, Tanzania, held in November 2001, that convened PLEC farmers, researchers, extension agents, policy-makers, and other stakeholders to discuss experiences and lessons from working with farmers in the field of agricultural biodiversity in East Africa. The time and commitment of researchers, extension agents, and other stakeholders from different institutions, universities, and departments in contributing to field and office work and their tireless and continuous visits and interactions with farmers are greatly appreciated.

The success of PLEC work in East Africa and the compilation of this book could not have been possible without the advice and encouragement of Professor Michael Stocking (PLEC associate scientific coordinator) throughout the six years of preparation and fieldwork. Clemmie Perowne, his research assistant at the University of East Anglia, provided excellent support to turn the Arusha proceedings into a book manuscript for UNU Press. Thanks also go to Beatrice E. Maganga in Mwanza, Tanzania, who compiled the first entry of the manuscript for the proceedings.

A word of thanks must go to Professor Harold Brookfield (principal scientific coordinator for PLEC) and Luohui Liang (managing coordinator of PLEC) for allowing and facilitating East Africa to have this book published. PLEC East Africa recognizes the PLEC executing agency support of the United Nations University (UNU), the financial support of the Global Environment Facility (GEF) in Washington, DC, and the GEF implementing agency support of the United Nations Environment Programme in Nairobi. It was only through these agencies that it became possible to create awareness among the international community of the importance of conserving and managing biodiversity in agricultural systems, and the positive experiences in East Africa. As the PLEC executing agency, the UNU managed cross-country activities and enabled the many and varied activities of the PLEC clusters, providing constant support, especially to PLEC East Africa. Without the continuous support of the Vice-Rector, Professor Suzuki, and the staff of the Environment and Sustainable Development Programme, especially including Masako Ebisawa (and her predecessor, Audrey Yuse), PLEC would not have worked and been the force it has been in East Africa.

Finally, it must be said that this is not strictly an academic book; it is a record of experiences and discoveries made by researchers and farmers. Many chapters have few, or no, references, for example. The farmer partners in PLEC East Africa have written of their feelings and reactions, not of some well-documented experimental design or closely argued rational arguments. Equally, PLEC's developing country research colleagues, many of whom were rather narrow natural resource scientists before PLEC, have discovered "the field". They have spent many hours, days, and weeks working with farmers, rather than listing literature references and undertaking scientific methodologies that would stand the scrutiny of most academic referees. They have appreciated the opportunity that GEF funding provided of engaging with land users, helping to set up demonstration sites, and facilitating the formation of user groups. In many ways these were development activities rather than research. However, through this close involvement the authors would claim that the research is far more targeted to issues of real interest to society, and especially to the farmers. The authors thank UNU Press for understanding the PLEC "voyage of discovery" by all its participants and stakeholders. It has given the authors of this book an opportunity to show to an international audience how and why it is wise to work with farmers, and sometimes to compromise academic integrity for the sake of gaining much richer knowledge of biodiversity, how to protect it, and what benefits such protection affords to human beings.

REFERENCES

Brookfield, H. 1995. "Postscript: The 'population-environment nexus' and PLEC", *Global Environmental Change*, Vol. 5, No. 4, pp. 381–393.

Brookfield, H. 2001. *Exploring Agrodiversity*. New York: Columbia University Press.

Brookfield, H. and C. Padoch. 1994. "Appreciating agrodiversity: A look at the dynamism and diversity of indigenous farming practice", *Environment*, Vol. 36, No. 5, pp. 6–11, 37–45.

Brookfield, H., C. Padoch, H. Parsons, and M. Stocking. 2002. *Cultivating Biodiversity: Understanding, Analysing and Using Agricultural Diversity*. London: ITDG Publishing.

Coffey, K. 2000. *PLEC Agrodiversity Database Manual*. New York: United Nations University, www.unu.edu/env/plec/.

Cromwell, E. 1999. *Agriculture, Biodiversity and Livelihoods: Issues and Entry Points*. London: Overseas Development Institute.

Kaihura, F., R. Kiome, M. Stocking, A. Tengberg, and J. Tumuhairwe. 1999. "Agrodiversity highlights in East Africa", *PLEC News and Views*, No. 14, pp. 25–32.

Konde, A., D. Dea, E. Jonfa, F. Folla, I. Scoones, K. Kena, T. Berhanu, and W. Tessema. 2001. "Creating gardens: The dynamics of soil fertility management in Wolayta, southern Ethiopia", in I. Scoones (ed.) *Dynamics and Diversity: Soil Fertility and Farming Livelihoods in Africa*. London: Earthscan, pp. 45–77.

Liang, L. 2002. "Promoting agrodiversity: The case of UNU project on people, land management and environmental change (PLEC)", *Global Environmental Change*, No. 12, pp. 325–330.

Liang, L., M. Stocking, H. Brookfield, and L. Jansky. 2001. "Biodiversity conservation through agrodiversity", *Global Environmental Change*, No. 11, pp. 97–101.

Rerkasem, K. 2002. *Final Report of PLEC-Thailand*. Chiang Mai: United Nations University.

Stocking, M. 2002. "Diversity: A new strategic direction for soil conservation", in *Sustainable Utilisation of Global Soil and Water Resources*, proceedings of the 12th International Soil Conservation Conference, 26–31 May 2002, Beijing. Volume 1. Beijing: Tsinghua University Press, pp. 53–58.

Stocking, M. and N. Murnaghan. 2000. *Handbook for the Field Assessment of Land Degradation*. London: Earthscan.

Tengberg, A., J. Ellis-Jones, R. M. Kiome, and M. Stocking. 1998. "Applying the concept of agrodiversity to indigenous soil and water conservation practices in eastern Kenya", *Agriculture, Ecosystems and Environment*, No. 70, pp. 259–272.

Tiffen, M., M. Mortimore, and F. Gichuki. 1994. *More People, Less Erosion: Environmental Recovery in Kenya*. Chichester: John Wiley.

Wood, D. and J. M. Lenné (eds). 1999. *Agrobiodiversity: Characterization, Utilization and Management*. New York: CABI, Wallingford and Oxford University Press.

2

Managing biodiversity in agricultural systems

Michael Stocking

Introduction

Biological diversity (or "biodiversity" for short) has always been a central concern of ecologists. While the topic itself derives from conservation biology (Primack 1998), the concern arises from different sources. First, there is claimed to be a current "mass extinction spasm", or the loss of tens of thousands of species over the past few decades, which is predicted to continue at an accelerated rate for the next few decades (Lawton and May 1995). Secondly, there are worries over what we are losing and whether this is something important for human society. One of the most pervasive potential damages is the loss of ecosystem function (Swift *et al.* 1996), with related effects on reduced productivity and lack of sustainability of agricultural systems. There are also moral, ethical, social, and economic arguments, but these are largely outside the scope of this chapter. Here, the principal issue is whether biodiversity is being preserved *in situ* and on farm, as an integral part of farming practice. In other words, do we have good examples of management of biodiversity, where the biodiversity itself is being retained for the possible benefit of future generations, and the biodiversity is at the same time supporting the livelihoods of local farmers? A further issue is to prove how addressing local issues of agrodiversity should be a concern of the international community. Or, how do we link the local with the global to demonstrate how both will benefit by a focus on agricultural biodiversity and its management?

Biodiversity refers to the variety, or the number, frequency, and variability, of living organisms and the ecosystems in which they occur. It includes species diversity, intra-species and genetic diversity, and ecosystem diversity. The loss of this biodiversity is a major concern of our time at a number of spatial and temporal scales. The concern can be distilled into four major areas: species extinction, and how that might prejudice future needs for genetic resources; the accelerating trends of losses, especially as land use moves to more marginal areas and intensifies in core areas; the synergistic effects of loss in biodiversity, such as the degradation of soils, depletion of fixed carbon stocks, and related effects on climate change; and our own survival. The standard reaction to these issues is to protect biodiversity in areas set aside for the specific purpose. For example, the IUCN (1994) has categorized areas for the management of biodiversity from I (strict nature reserves and wilderness areas) through to VI (managed resource protected areas). This last category is the only one to admit the sustainable use of natural resources by people, but under strict controls that prioritize the preservation of biological diversity. Is this approach to the management of biodiversity itself sustainable?

Many would now question the singular approach towards protected areas as the basis for managing the world's biodiversity. With an eye very much on the situation in East Africa, David Western (1989: 158) commented that "if we cannot save nature outside protected areas, not much will survive inside". East Africa has a reasonably good record of establishing protected areas (approximately 9 per cent of land area, including renowned reserves such as the Serengeti (Tanzania), Queen Elizabeth (Uganda), and Tsavo (Kenya)). But Africa as a whole claims only 4.6 per cent of land as protected and 2.5 per cent managed (IUCN Category VI) according to WRI (1994), which is significantly below world averages. Therefore, it has to be recognized that most of the land in most countries will never be protected in this way. Additionally, relying on parks and reserves to protect biological diversity creates a "siege mentality" (McNeely 1989) or the "fortress conservation" and "fences and fines" approach (Wells, Brandon, and Hannah 1992). It is a compelling narrative that has been particularly pervasive in sub-Saharan Africa. However, as Adams and Hulme (2001) describe, the fortress has been attacked, especially in the area of wildlife conservation, but also more generally in community-based approaches to conservation that include soils, lands, forests, and water. The main dimension of this counter-narrative, appropriate to this book, is the linking of conservation objectives to local development needs. Examples include "conservation-with-development" projects, as in the Usambara Mountains of Tanzania (Stocking and Perkin 1992), where protection of the endemic species was linked to providing for sustainable use of forest resources and benefit-sharing for protection.

The community conservation narrative

The community conservation narrative is now well established, and the PLEC project is emerging as part of it and of mainstream conservation. It has been a success because it arose "at a time of significant shifts in the dominant discourses of development". (Adams and Hulme 2001: 17). The 1970s had been a decade of "top-down" technocratic approaches to development, but there was considerable questioning as to their effectiveness. They failed to deliver the promised economic growth and social benefits. So the 1980s saw a reversal of the process of development to "bottom-up planning", "participation", "gender awareness", and "empowerment of local people". Some of this was simply rhetoric. However, during the 1990s the tools to deliver the reversal were developed and became more widely disseminated. Participatory rural appraisal and its techniques became standard practice, and this helped to forge the link between those who wanted conservation and those who wanted social development. It was a strong incentive for conservationists to become involved more with people than with animals or natural forests. It also released significant sources of funds to conservation that would otherwise have been devoted solely to development aid.

The global PLEC project, some of whose history is related in Chapter 1, was a child of these changes in narrative towards community conservation, understanding local people's perspectives, and devoting resources to development that at the same time promoted conservation. Indeed, PLEC in the early 1990s was devised as a project to examine the "population-environment nexus", or the claims that increasing populations cause increasing demands on natural resources and hence environmental damage. This "nexus" hypothesis was a pervasive part of the "fortress conservation" narrative. Deal with population growth, prevent people from invading pristine natural habitats, and control land use, then conservation of biodiversity and other natural resources could be achieved. In *PLEC News and Views* No. 1, July 1993, the scientific coordinator said: "*PLEC* emphasises the consequences for land management, and hence for the environment, of continuing population growth" (Brookfield 1993: 2). Agenda 21 arising from the 1992 Rio Earth Summit was entirely consistent with this argument, quoting alarming figures of resource depletion with predictions of global populations of 8.5 billion by

2025. However, by the mid-1990s PLEC was moving rapidly sideways to embrace an alternative hypothesis that conservation could be achieved by working with local people to meet their legitimate needs. Indeed, evidence was starting to accumulate that population growth could in some circumstances be seen as a factor for environmental protection rather than a cause of environmental degradation – the so-called Boserupian hypothesis that population increase induces innovation and technological adaptation. So by 1996 PLEC was saying that "the most important aim ... is to undertake participatory planning and experimental work with communities, with the purposes of protecting biological diversity and improving agricultural sustainability and water management" (Guo, Dao, and Brookfield 1996: 14). This focus on community, households, working with farmers, understanding local experts, utilizing indigenous knowledge, and participatory experimentation has remained with PLEC to today. As will be demonstrated later in this chapter, PLEC is about how biodiversity is managed by local people, and what lessons we can draw from examples of "good practice". East Africa PLEC has been at the fore in accepting the new narrative and undertaking research with local people as the key stakeholders.

The Convention on Biological Diversity

In 1992, at the United Nations Conference on Environment and Development in Rio de Janeiro, Brazil, two binding global agreements were signed. These were the Convention on Climate Change, which targets industrial and other emissions of greenhouse gases such as carbon dioxide, and the Convention on Biological Diversity (CBD), the first global agreement on the conservation and sustainable use of biological diversity (see Table 2.1). The biodiversity treaty gained rapid and widespread ac-

Table 2.1 The origin and history of the Convention on Biological Diversity

- *1987: UNEP Governing Council* – working group to harmonize existing efforts in biodiversity
- *May 1989: UNEP Expert Working Group* – prepare an international legal instrument for conservation and sustainable use of biodiversity
- *February 1991: Intergovernmental Negotiating Committee* – adoption of Nairobi Final Act of the Conference for the adoption of the agreed text of the convention
- *June 1992: UNCED, Rio* – CBD opened for signature
- *29 December 1993: entry into force* – 50 signatories
- *By Second GEF Assembly, Beijing, October 2002* – 186 parties, but notable absentees (Thailand and the USA, for example)

ceptance. Over 150 governments signed the document at the Rio conference, and today there are 186 parties to the convention and 168 national
signatories – see the convention website: www.biodiv.org.

The convention has three main objectives:
- the conservation of biodiversity
- sustainable use of the components of biodiversity
- sharing the benefits arising from the commercial and other utilization
  of genetic resources in a fair and equitable way.

Therefore, out of three objectives, two are very much targeted at human needs and developmental goals. The CBD is comprehensive. It recognizes for the first time that the conservation of biological diversity is "a
common concern of humankind" and is an integral part of the development process. The agreement covers all ecosystems, species, and genetic
resources. It links traditional conservation efforts to the economic goal of
using biological resources sustainably. It sets principles for the fair and
equitable sharing of the benefits arising from the use of genetic resources,
notably those destined for commercial use. It also covers the rapidly expanding field of biotechnology, addressing technology development and
transfer, benefit-sharing, and biosafety. Importantly, the CBD is legally
binding; countries that join it are obliged to implement its provisions. To
assist in implementation, the Global Environment Facility (GEF) fund
was established where developed countries pledged finance to secure developing countries' compliance with the responsibilities of being a convention signatory. To date, some US\$6 billion has been assigned to the
GEF, and the third replenishment endorsed at the GEF Assembly in
Beijing in October 2002 promised a further US\$2.6 billion for 2003–2006.

The CBD reminds decision-makers that natural resources are not infi-
nite, and sets out a new philosophy for the twenty-first century – that
of sustainable use. While past conservation efforts were aimed at protecting particular species and habitats, the CBD recognizes that ecosystems, species, and genes must be used for the benefit of humans. However, this should be done in a way and at a rate that does not lead to the
long-term decline of biological diversity. The CBD also offers decision-
makers guidance based on the precautionary principle that where there is
a threat of significant reduction or loss of biological diversity, lack of full
scientific certainty should not be used as a reason for postponing measures to avoid or minimize such a threat. The CBD acknowledges that
substantial investments are required to conserve biological diversity. It
argues, however, that conservation will bring us significant environmental, economic, and social benefits in return.

The funding of convention responsibilities through the GEF is organized firstly through an operational strategy (see Table 2.2) and then

Table 2.2 Extracts from the GEF operational strategy for biological diversity

- "Biodiversity is a source of significant economic, aesthetic, health and cultural benefits which form the foundation for sustainable development"
- "Loss of biodiversity poses a global threat to human well-being"
- "Biodiversity is not equally distributed"
- "Adoption of the *CBD* [is recognition of] the intrinsic value of biodiversity and ... its sustenance of life support systems of the biosphere"

through a set of operational programmes (OPs). In the case of biological diversity, these OPs consist of four eco-regional programmes (arid, aquatic, forest, and mountain), plus a relatively new OP Number 13 to cover "conservation and sustainable use of biological diversity important to agriculture". It is this last that is now the most relevant to spearheading the management of biodiversity and supporting projects to achieve the two objectives of the CBD that directly relate to human needs. As cited in OP13 (see website: http://gefweb.org/Operational_Policies/Operational_Programs/OP_13_English.pdf), activities are aimed at sustaining the functions of biodiversity in agricultural ecosystems. This is, in turn, to maintain and enhance the goods and services provided by biodiversity, including both those which support agricultural production and wider services such as provision of clean water, control of soil erosion, and moderation of climatic effects. It is intended that the impact of agriculture would be integrated into the planning and management of the wider ecosystem. Although the approval of PLEC pre-dated OP13 by at least two years, this focus on agricultural biodiversity has helped to highlight the crucial role of managed biodiversity, and the management practices land users employ both to protect their biodiversity and to secure their livelihoods.

The PLEC agrodiversity framework

PLEC has used, since near its inception, a framework that brings together the principal components of how biodiversity is managed on agricultural lands (Figure 2.1). Agrodiversity is divided into four principal elements, all of which overlap but each of which constitutes distinctive elements that have their own rationale and means of assessment.

These categories are fundamental to understanding the interface between natural biological diversity and human land use. They operate through a variety of spatial scales, which are important to distinguish. Typically biological diversity is situated as an attribute of an organism

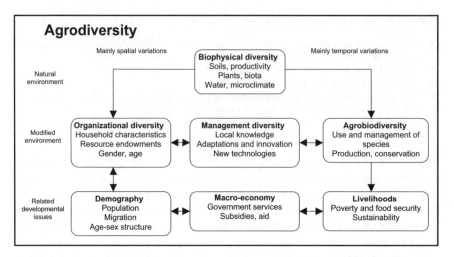

Figure 2.1 Elements of agrodiversity – main components and principal development issues
Source: Brookfield and Stocking (1995)

or a site. Agrodiversity is not only plant and site specific, but it is also a feature of whole fields, farms, communities, and landscapes. At all these spatial scales, land users manage the biodiversity, thereby constructing their livelihoods while at the same time protecting a form of biodiversity that is useful to society.

Biophysical diversity

This is biological diversity closest to that found in the natural environment, which controls (especially in low-input farming systems) the intrinsic quality of the natural resource base that is utilized for production. It contains the natural resilience of the biophysical environment, to be exploited by agricultural systems. It therefore includes soil characteristics and their productivity, and the biodiversity of natural (or spontaneous) plant life and of the soil biota. It takes account of both physical and chemical aspects of the soil, surface and near-surface physical and biological processes, hydrology, microclimate, and also variability and variation in all these elements. Farmers select within this diversity and they often manipulate it quite substantially. Sometimes this management goes to the extent of "manufacturing" soils and remodelling the landscape in, for example, terracing. Created biophysically diverse areas might include protected areas and nature reserves.

Management diversity

In addition to land-transforming management as mentioned above, management diversity also'includes all methods of managing the land, water, and biota for crop production and the maintenance of soil fertility and structure. Included are biological, chemical, and physical methods of management. Management may not only be specific to certain soils and terrains, but also to different seasons. Smallholder farmers are often adept at altering practices according to rainfall patterns. Some biological management, such as the reservation of forest for watershed protection or the planting of live hedges, has direct physical consequences. Local knowledge, constantly modified by new information, is the foundation of this management diversity.

Agrobiodiversity

Agrobiodiversity embraces all crops and other plants used by or useful to people and, by also involving biota having only indirect value to people, it cannot be sharply distinguished from total plant biodiversity. Particularly important is the diversity of crop combinations, and the manner in which these are used to sustain or increase production, reduce risk, and enhance conservation.

Organizational diversity

Often called the "socio-economic aspects", organizational diversity includes diversity in the manner in which farms are owned and operated, and in the use of resource endowments. It underpins and helps explain "management diversity" and its variation between particular farms. Explanatory elements include labour, household size, the differing resource endowments of households, and reliance on off-farm employment. Also included are age group and gender relations in farm work, dependence on the farm as against external sources of support, the spatial distribution of the farm, and differentials between farmers in access to land.

Importance of biodiversity to local communities

Biodiversity is, above all, important to the livelihoods of local communities. The relationship is not necessarily simple, and can sometimes be negative. Biodiversity loss may, therefore, have ambivalent effects on livelihoods according to the nature of the productive enterprises introduced in place of natural vegetation. A decline in livelihoods may arise

Table 2.3 Links between biodiversity and local livelihoods

	Decline in livelihoods	Improvement in livelihoods
Biodiversity loss	**Intensive and large-scale resource extraction:** e.g. – loss of steep land forests causing erosion and loss of natural capital – surface mineral extraction destroying vegetation and removing land from local people	**Conversion of natural habitats to agriculture or forestry:** e.g. – commercial land-use systems and mono-cultures, such as tea estates on hill lands; enables extraction of food and products to lowland economies – commercial forests; gain in employment possibilities and added-value products for local people
Biodiversity maintenance or increase	**Strict protected areas:** e.g. – conservation benefits gained, but access to local communities denied – predation by wild animals may increase	**Sustainable management of biodiversity:** e.g. – poor and marginal communities that depend upon biodiversity – they protect it, manage it, and utilize the products, often with a view to future generations – society and culture enhanced by protection of biodiversity

Source: Adapted from IUCN (2000)

because species important to local people are no longer available. Table 2.3 sets out a matrix of the possible permutations of the link between biodiversity and livelihoods for smallholder farmers on marginal lands. Loss of biological resources such as timber species or non-timber products such as medicinal herbs has implications for not only natural capital but also social and human capital. This wholly extractive and usually externally controlled transforming process has extremely deleterious consequences for livelihoods that are perhaps only very modestly compensated for by temporary labour opportunities as forest workers. However, as the PLEC project is demonstrating, a loss of natural biodiversity may be translated into an increase in "agrodiversity".

The conversion of natural habitats may go two ways: to monocultures and loss in crop genetic diversity; or to complex small-scale agricultural

systems that maintain or increase overall biodiversity. The key for interventions is to ensure support for the win-win scenarios (lower right-hand
box of Table 2.3 – darker shading), avoid the lose-lose scenarios (lighter
shading) at all costs, and attempt to convert the other changes to gain net
benefit. Protected areas, for example, under the CBD should attempt to
share the benefits with local people through tourist revenues or local
rights of access to sustainable harvesting of timber. Similarly, conversion
of natural vegetation to agriculture can be accompanied by an increase in
biodiversity rather than a loss if production and conservation goals are
targeted simultaneously.

Local people are, therefore, the guardians of local biodiversity. They
are also the motor for protection, since biodiversity underwrites their
livelihoods. They derive part of their livelihood from biodiversity in the
form of food, fuel, fodder, and shelter. They gain security in diversity,
enabling them to cope with external forces such as climatic variability
or market forces. Biodiversity, in effect, underwrites sustainability. The
"best practice" situation of biodiversity increased and livelihoods improved as promoted by PLEC is one that demands management. It is
therefore a managed biodiversity, often consisting of complex agroforestry systems and multistorey cropping. On the Mount Meru demonstration site in Tanzania, coffee is grown under shade trees, interspersed with
many banana varieties. In the Bushwere demonstration site, Uganda,
complex groupings of different varietals of bananas are maintained in
production by high-quality mulching materials, often brought from the
upper slopes. In Embu, Kenya, farmers have an intimate knowledge of
the benefits of associations of crops and trees, such as *Ficus* spp. So
farmers convert elements of the natural biodiversity, domesticate some
species, and then add a huge range of other species and varieties from
elsewhere.

Managing the soil

To illustrate the management of biodiversity, consider the role of soil in
the managed agro-ecosystem. Soil is a dynamic medium, responsive to
natural environmental variability, to how it is managed and improved
in terms of fertility, and to the way activities on the farm are organized.
It has its own below-ground biodiversity, which makes nutrients available and undertakes a complex of ecosystem service functions. Figure 2.2
turns the agrodiversity framework of Figure 2.1 around to focus on soil
agrodiversity and its management.

In practical terms, soil agrodiversity involves the inherent soil fertility,
changes in soil quality brought about by external forces, soil conservation

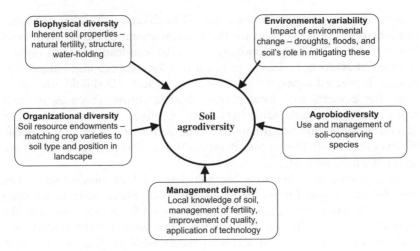

Figure 2.2 A focus on aspects of the soil and its management

Table 2.4 Examples of soil agrodiversity from East Africa

- Siting of banana cultivars in SW Uganda and support to rural livelihood
- Resilience of mixed farming systems, Ng'iresi village, Tanzania
- *Ngoro* pits of SW Tanzania
- Keeping certain trees in fields for soil improvement, Embu, Kenya
- Multi-purpose stone walls in fields, Mbarara, Uganda

techniques, local knowledge of soil management, and the value of soil as an economic and cultural resource. By recognizing soil agrodiversity, beneficial attributes may be enhanced: there are site benefits in the form of increased soil resilience and ability to withstand forces such as water erosion; management and organizational benefits, through promoting practices that conserve the soil and increase production; and landscape and social benefits, through, for example, diversifying the local economy. These are complex and often indirect linkages, but all have important implications for biodiversity and sustainable rural livelihoods (see Table 2.4).

Figure 2.3 brings together the experience from East Africa of soil agrodiversity and its management. There are a large number of soil practices and techniques that farmers have learnt to apply in different situations, and probably only a small percentage are captured in Figure 2.3. However, all have implications for biodiversity and livelihoods. Without the active and knowledgeable management of biodiversity by smallholder farmers, the many benefits of biodiversity could not be realized.

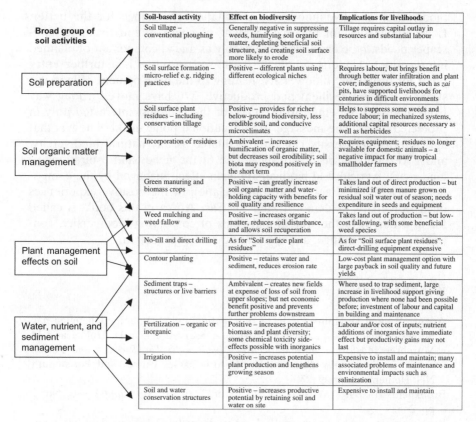

Broad group of soil activities

- Soil preparation
- Soil organic matter management
- Plant management effects on soil
- Water, nutrient, and sediment management

Soil-based activity	Effect on biodiversity	Implications for livelihoods
Soil tillage – conventional ploughing	Generally negative in suppressing weeds, humifying soil organic matter, depleting beneficial soil structure, and creating soil surface more likely to erode	Tillage requires capital outlay in resources and substantial labour
Soil surface formation – micro-relief e.g. ridging practices	Positive – different plants using different ecological niches	Requires labour, but brings benefit through better water infiltration and plant cover; indigenous systems, such as *zai* pits, have supported livelihoods for centuries in difficult environments
Soil surface plant residues – including conservation tillage	Positive – provides for richer below-ground biodiversity, less erodible soil, and conducive microclimates	Helps to suppress some weeds and reduce labour; in mechanized systems, additional capital resources necessary as well as herbicides
Incorporation of residues	Ambivalent – increases humification of organic matter, but decreases soil erodibility; soil biota may respond positively in the short term	Requires equipment; residues no longer available for domestic animals – a negative impact for many tropical smallholder farmers
Green manuring and biomass crops	Positive – can greatly increase soil organic matter and water-holding capacity with benefits for soil quality and resilience	Takes land out of direct production – but minimized if green manure grown on residual soil water out of season; needs expenditure in seeds and equipment
Weed mulching and weed fallow	Positive – increases organic matter, reduces soil disturbance, and allows soil recuperation	Takes land out of production – but low-cost fallowing, with some beneficial weed species
No-till and direct drilling	As for "Soil surface plant residues"	As for "Soil surface plant residues"; direct-drilling equipment expensive
Contour planting	Positive – retains water and sediment, reduces erosion rate	Low-cost plant management option with large payback in soil quality and future yields
Sediment traps – structures or live barriers	Ambivalent – creates new fields at expense of loss of soil from upper slopes; but net economic benefit positive and prevents further problems downstream	Where used to trap sediment, large increase in livelihood support giving production where none had been possible before; investment of labour and capital in building and maintenance
Fertilization – organic or inorganic	Positive – increases potential biomass and plant diversity; some chemical toxicity side-effects possible with inorganics	Labour and/or cost of inputs; nutrient additions of inorganics have immediate effect but productivity gains may not last
Irrigation	Positive – increases potential plant production and lengthens growing season	Expensive to install and maintain; many associated problems of maintenance and environmental impacts such as salinization
Soil and water conservation structures	Positive – increases productive potential by retaining soil and water on site	Expensive to install and maintain

Figure 2.3 Soil agrodiversity – components, implications, and effects

Conclusion

People build livelihoods in smallholder environments by managing the biological diversity at their disposal. PLEC, especially in East Africa, has demonstrated in this book and its other outputs that management is the key to achieving biodiversity conservation on agricultural lands. There is a wealth of innovation, creation, and knowledge on smallholder farms, leading to conservation of biodiversity, sharing of its benefits, and the long-term sustainability of the natural resources – the three objectives of the Convention on Biological Diversity.

There are important lessons arising from our understanding of how local societies cope with their difficult biophysical environments. Biodiversity conservation need not run counter to the needs and interests of local people. Interventions need to be targeted on areas where both biodiversity and livelihoods can simultaneously change for the better. Local knowledge on the management of complex and dynamic landscapes needs to be employed, preferably as a lead component of planned interventions. Understanding the basic soil resource is a further entry point for protecting biodiversity, leading to the development of land-use systems that are resilient and productive. We have moved a long way from biodiversity being the preserve of the ecologist interested only in protected areas. The message to international policy-makers now is that agricultural biodiversity can be as valuable as any natural biodiversity, and is certainly more obviously linked to the needs and aspirations of land users. Agricultural biodiversity is, in effect, managed by the components of agrodiversity. Greater appreciation of the positive experiences of agrodiversity, as demonstrated for East Africa in this book, is called for in order to further the global goals of conserving biodiversity, controlling climate change, and preventing land degradation.

REFERENCES

Adams, W. and D. Hulme. 2001. "Conservation and community: Changing narratives, policies and practices in African conservation", in D. Hulme and M. Murphree (eds) *African Wildlife and Livelihoods: The Promise and Performance of Community Conservation*. Oxford: James Currey and Portsmouth, NH: Heinemann, pp. 9–23.

Brookfield, H. 1993. "What is *PLEC* about?", *PLEC News and Views*, No. 1, pp. 2–8.

Brookfield, H. and M. Stocking. 1995. "Agrodiversity: Definition, description and design", *Global Environmental Change*, No. 9, pp. 77–80.

Guo, H., Z. Dao, and H. Brookfield. 1996. "Agrodiversity and biodiversity on the ground and among the people: Methodology from Yunnan", *PLEC News and Views*, No. 6, pp. 14–22.

IUCN. 1994. *Guidelines for Protected Area Management*. Gland, Switzerland: International Union for the Conservation of Nature and Natural Resources.

IUCN. 2000. "Biodiversity in development: The links between biodiversity and poverty", *Biodiversity Brief 1*. Gland: World Conservation Union, with the European Commission, Brussels and the Department for International Development, London, http://wcpa.iucn,org/pubs/pdfs/biodiversity/biodiv_brf_01.pdf.

Lawton, J. H. and R. M. May (eds). 1995. *Extinction Rates*. Oxford: Oxford University Press.

McNeely, J. A. 1989. "Protected areas and human ecology: How national parks can contribute to sustaining societies of the twenty-first century", in D. Western and M. Pearl (eds) *Conservation for the Twenty-first Century*. Oxford: Oxford University Press, pp. 150–157.

Primack, R. B. 1998. *Essentials of Conservation Biology*, 2nd edn. Sunderland, MA: Sinauer Associates.

Stocking, M. and S. Perkin. 1992. "Conservation-with-development: An application of the concept in the Usambara Mountains, Tanzania", *Transactions of the Institute of British Geographers*, No. 17, pp. 337–349.

Swift, M. J., J. Vandermeer, P. S. Ramakrishnan, J. M. Anderson, C. K. Ong, and B. A. Hawkins. 1996. "Biodiversity and agroecosystem function", in H. A. Mooney, et al (eds) *Functional Roles of Biodiversity: A Global Perspective*. Chichester: John Wiley, pp. 261–298.

Wells, M., K. Brandon, and L. Hannah. 1992. *People and Parks: Linking Protected Areas with Local Communities*. Washington, DC: World Bank.

Western, D. 1989. "Conservation without parks: Wildlife in the rural landscape", in D. Western and M. Pearl (eds) *Conservation for the Twenty-first Century*. Oxford: Oxford University Press, pp. 158–165.

WRI. 1994. *World Resources 1994–95: A Guide to the Global Environment*. Washington, DC: Oxford University Press for the World Resources Institute.

3

Agrobiodiversity potential of smallholder farms in a dissected highland plateau of western Uganda

Joy Tumuhairwe and Charles Nkwiine

Introduction

Agrobiodiversity is a fundamental component of biodiversity and of particular importance in Uganda, where 21 per cent of total land area is under agricultural use and 43 per cent of the national gross domestic product (GDP) comes from agriculture. While humans depend on agrobiodiversity for food, medicine, and industrial uses, other biodiversity units (such as forests and gorilla parks) are protected because they are considered aesthetically valuable. Only a properly functioning and diverse land use provides the key ecosystem services – those of supplying air, water, and soil.

Uganda's hilly and mountainous areas have been globally designated a "centre of plant diversity" by the International Union for the Conservation of Nature (IUCN) plant conservation programme because of their high number of "Afromontane plant species". By definition, these places are considered particularly rich in plant life (Heywood and Davies 1996), which if adequately protected would assure the survival of the majority of the world's wild plants. The rich natural plant biodiversity of these montane ecosystems is also reflected in the large number of different crops and land-use types that these areas are able to support because of the productivity of the soils, vegetation, and other land resources. These areas include Bundibugyo, Bushenyi, Kabale, Kabarole, Kapchorwa, Karamoja, Kasese, Kisoro, Mbale, southern parts of Mbarara district, much of Ntungamo, and Rukungiri. This is over 20 per cent of all districts of the country. Although they have extremely steep terrain, the fertile soils and conducive climate (bimodal and high rainfall) have attracted farming communities. In addition to the inherently rich bio-diversity resources – some of which are of global significance – these agro-ecosystems are well known for their contribution to national food security and household income, mainly from agrobiodiversity products. The people in these areas are traditionally cultivators and polygamous in lifestyle. Populations are growing rapidly (3 per cent per annum) and land clearing for farming is widespread. Population densities range from 200 to 700 persons per square kilometre (Statistics Department 1992). Pressures brought on by agriculture and population growth in these steeply sloping lands reduce vegetative cover of soil surface, destroy soil structure, and expose the inherently friable soils to the strong and desic-cating winds characteristic of such mountainous regions. All these render the ecosystem fragile and thus vulnerable to degradation.

Due to the fragility of these important agro-ecosystems that have been exposed to intensive agriculture, there is a need to reduce and/or halt loss of biodiversity resources if the ecosystems are to remain functionally sustainable. This will require integrating biodiversity conservation efforts into the farming practices of smallholders who are the daily managers of the resources. In this chapter, PLEC-Uganda gives its experiences in try-ing to promote this integration with the aim of identifying, developing, and promoting acceptable technologies for integrating biodiversity con-servation into smallholder farm units with a net benefit of improved household income and welfare. The objectives of the study were to estab-lish the status of agrobiodiversity, and to work with smallholder farmers to develop sustainable technologies for biodiversity conservation on ag-ricultural lands on rugged highlands.

Background

The study area is in Mbarara district, south-western Uganda. It consists of two counties – Rwampara and Isingiro – both located south of Mbara town and approximately 30 km north of the Tanzanian border. It borders with Ntungamo district in the west and Lake Mburo National Park in the east. The area consists of two major agro-ecological zones: one of pasto-ral, dry to semi-arid rangelands in Isingiro county (east – termed zone A) and the other a grassland zone with mixed farming systems changing to high-altitude Afromontane in Rwampara (west – zone B). Both zones are important ecosystems of the East African region that are valuable for food security and rural livelihoods. However, their resources are

rapidly being degraded due to demographic and socio-economic transformations.

The terrain is rugged, with altitude ranging from 1,400 to 1,800 metres
above sea level. The area is characterized by high plateaus and dissected
by ravines that lead to narrow U-shaped valleys which open to wider
valleys and plains in the east. Populations have increased from about 80
to 250 people per square kilometre in Mwizi subcounty of Rwampara and
10 to 100 people per square kilometre in Kabingo Isingiro between 1959
and 1991 (Statistics Department 1992).

The following participatory methods were used throughout the study.

Transects

A mega-transect of 5 km × 30 km was established, stretching from Ru-
bingo parish in Bugamba to Kamuli parish in Kabingo. Activities included:
* drive transects to identify the major land-use systems
* community workshops
* eight line-and-belt transects of 2 km cutting across different land uses
  that are representative of particular areas within the mega-transect.

Demonstration site

A simple scoring system was devised to select demonstration sites from
17 transected parishes (10 and seven from agro-ecological zones A and
B respectively). Selection involved noting the agro-ecological zone and
then scoring the parishes according to five criteria:
* receptiveness of the people (R)
* ethnic diversity (E)
* accessibility (AC)
* number of land-use types (L)
* number of crop combinations (C).

Meanwhile, selection of sample areas and plots was based on variations
in field types; cooperation of field owners; and replication and spread
over different villages of the parish. Replication across the landscape
types was only possible for a few crop combinations because some field
types occur on specific locations on the landscape. An inventory of plant
diversity was taken in accordance with guidelines from the PLEC biodiversity advisory group (Zarin, Guo, and Enu-Kwesi 2002).

Farmers to be involved in the demonstrations were initially identified
by fellow farmers. These were then confirmed or revised later by the
PLEC scientists who visited the individual candidates' fields to ascertain
the following basic criteria:

- innovation in conserving several plant species or varieties in the cropping system
- innovation in management of the system, including spatial arrangement, soil management, timeliness in planting, weeding, and other crop agronomic aspects
- degree of understanding and explanation of techniques
- willingness to seek or take up more information and skills
- ability to learn, work with PLEC scientists, and change where necessary
- willingness to demonstrate and train other farmers and other stakeholders.

Participatory biodiversity inventories in sample areas helped to familiarize collaborating scientists and farmers with the demonstration site and farmers' characteristics, and to refine the selection of expert farmers and demonstration activities. Farmers who qualified were referred to as "expert farmers" and were facilitated to demonstrate sustainable methods of agrobiodiversity conservation.

Demonstration activities

Demonstration activities involved a variety of projects.
- Participatory evaluation of expert farmers' innovations.
- Regular field visits of candidate farmers and activities to exchange knowledge, experiences, and ideas.
- Farmer experimentation of models or their components that required testing.
- Adoption of the necessary improvements by expert farmers.
- Demonstrations to other farmers, local leaders, and other stakeholders during field workshops.
- Field visits by/to other collaborating farmers to share experiences and knowledge. Selection of participants for the farmer field exchange visits and demonstrations was made by the hosting farmer alongside the field extension worker and PLEC scientists. Emphasis was on fair representation of farmers from each village and also inclusion of at least one official village leader and some collaborating farmers in each case.

Dissemination

Dissemination of innovative approaches was through:
- farmer-to-farmer field visits, either individually or in groups
- field training sessions led by expert farmers with the PLEC scientists providing technical and logistical back-up
- field evaluation of developing technologies carried out by separate

groups of farmers, local leaders, and district-level experts in agriculture, environment, forestry, and community development.

Sustainability

Methods used to ensure the sustainability of PLEC initiatives included the following.
- Motivating expert farmers.
- Participatory assessment and evaluation.
- Involvement of stakeholders (namely farmers, members of the community, local leaders, and policy-makers at local and national levels).
- Strengthening common-interest farmer groups around the expert farmers, focusing on sustainable agrobiodiversity conservation as a major objective. The PLEC scientists facilitated the groups with technical guidance and logistical support in obtaining official registration and banking facilities as well as translating and typesetting their constitutions, logos, letterheads, and initial project proposals. The groups defined their own membership objectives and activities and managed their programmes accordingly.
- Development of policy and technical recommendations (see Chapter 21 in this volume).

Status of agrobiodiversity

Reconnaissance surveys of the mega-transect led to the identification of eight main land-use systems, as shown in Table 3.1. They consisted mainly of grassland-based systems (livestock), perennial crop-based systems, annual crops-based systems, and integrated livestock and perennial/annual crops-based systems.

Twenty-five major land uses were encountered on the eight line transects, as summarized in Table 3.2. Recognizing the influence of landscape diversity on land use and biodiversity, the 25 land uses were recorded against landscape types. These landscape types were defined according to position: hilltop (HT), backslope (BS), shoulder (SH), and footslope (FS).

The landscape position with the most varied land use was backslopes (23 land-use types or LUTs), followed by footslopes (21), hilltops (16), and shoulders (15). Farmers attributed the higher number of LUTs on backslopes to the many ecological stresses such as shallow soils, steep slopes, drying winds, and high losses of soil and moisture which induce them to develop a large diversity of land uses. Due to land shortage, farmers try to grow all types of crops and crop combinations on the mar-

Table 3.1 Land-use systems of southern Mbarara and their main characteristics

Transect name	Parish	Main characteristics
1. Rubingo	Rweibogo	Banana/coffee/cattle system. Settlements in valley and footslopes.
2. Bushwere	Bushwere	Intensive annual cropping, with intercropping and scattered small banana plantations. Few coffee fields mostly intercropped with bananas. Cultivation and settlements on all landscape types.
3. Ngoma	Ngoma	Intensive cropping with banana and annuals in approximately equal proportions. Some livestock and woodlots. Settlements mostly on ridge tops. Relatively good banana management.
4. Kashojwa	Rukarabo	Expansive banana plantations in valleys, annuals on hill slopes and tops. Grasslands on steep backslopes. Settlements variable for different hills.
5. Kigaaga	Kigaaga	Predominantly annuals on all landscape types. Some bushes in valleys, few bananas, and more pure-stand cropping practices.
6. Butenga	Kisuro	Mostly grasslands in wide valleys and plains. Paddocked pastures, much livestock, and little cropping.
7. Kagando	Kamuri	Grasslands and scrublands with fairly large banana plantations in valleys, uncultivable slopes with poor grasses, and widespread bush-burning practices.
8. Byaruha	Nyakigyera	Annual crop on hilltops and pediment, steep uncultivable slopes with grasslands and extensive banana plantations in both narrow and wide valleys. Settlements in footslopes.

ginal backslopes. They diversify their risks by growing many types of crops as an insurance against crop failure and avoidance of total economic loss. Unlike the hilltops and shoulders, which are lateritic and thus have shallow soils, the backslopes and footslopes have deep and loamy soils and can therefore support many different plants. However, there is a problem of lack of appropriate soil conservation measures.

While banana, maize, Irish potatoes, and finger millet are grown on all landscape types, banana growing dominates the valleys and ravines and is rare on the shoulder areas due to the shallowness of the soil. Peas are mostly grown on ridge tops and shoulders, while cassava is limited to the

Table 3.2 Distribution of major land uses along the landscape positions

Major land uses	% occurrence per landscape position				
	HT	SH	BS	FS	Total
Eucalyptus	45.5	–	36.4	18.2	100
Sorghum	21.4	14.3	35.7	28.6	100
Banana	12.0	7.2	38.6	42.2	100
Finger millet	31.8	18.2	31.8	18.2	100
Maize	41.7	29.2	20.8	8.3	100
Coffee	7.7	–	61.5	30.8	100
Sweet potatoes	30.8	30.8	30.8	7.7	100
Fallow	16.1	32.1	35.7	16.1	100
Pasture	18.2	9.1	36.4	36.4	100
Peas	30.0	60.0	–	10.0	100
Beans	21.1	21.1	26.3	31.6	100
Grassland	1.9	18.5	53.7	25.9	100
Settlement	15.4	23.1	38.5	23.1	100
Sugarcane	–	–	100.0	–	100
Bushland	7.7	19.2	38.5	34.6	100
Cassava	–	–	80.0	20.0	100
Natural forest	–	–	100.0	–	100
Irish potatoes	13.3	60.0	20.0	20.0	100
Road	12.5	13.3	13.3	60.0	100
Groundnuts	–	25.0	37.5	25.0	100
Soybean	–	–	100.0	–	100
Gully	–	–	33.3	66.7	100
Pineapple	–	–	100.0	–	100
Kraal	–	–	50.0	50.0	100
Wetlands	–	–	–	100.0	100
Total land uses	**16**	**15**	**23**	**21**	

lower slopes (BS and FS). Coffee – being perennial and deep-rooted – is also mostly grown on backslopes and footslopes.

Growing more than one crop in the same field promotes agrobiodiversity conservation. Several crop combinations were recorded in the study area. It is important to note that Bushwere had the highest number of crop combinations (four) or intercrops (16) compared to the others (four to eight). At the same time, Bushwere and Ngoma also had more pure-stand fields (seven) than the other transects, which had only four to six. These results guided the selection of demonstration sites from the 17 parishes.

Table 3.3 shows scores for each parish according the criteria. Bushwere had the highest score (28), followed by Kamuri and Nyakigyera (24 each). Consequently, Bushwere was taken up as a PLEC demonstration site for Uganda.

Table 3.3 Parish scores by criteria for selection of a demonstration site

Parish	(Agro)	(R)	(E)	(Ac)	(L)	(C)	Total
Rweibogo	A	3	2	1	3	5	13
Kabarama	A	3	2	1	3	5	14
Bushwere*	A	3	4	2	3	16	28
Kigaaga	A	3	2	1	3	6	15
Rwamiyonga	A	3	2	1	3	6	15
Rukarabo	A	3	4	3	3	8	21
Ngoma	A	2	4	2	4	8	20
Ibumba	A	3	4	2	2	6	17
Kigyendwa	A	3	4	2	2	5	16
Nyamuyanja	A	1	4	2	2	5	14
Katanoga	B	2	7	1	2	5	17
Kisuro	B	2	7	1	1	3	14
Nyakigyera	B	3	7	1	4	8	23
Kaharo	B	3	7	3	2	4	19
Katembe	B	3	7	2	2	3	17
Kagarama	B	3	8	2	3	3	19
Kamuri*	B	3	8	4	5	4	24

Note: The higher the scores, the greater the agrobiodiversity and reason for choice as a demonstration site.
* Indicates the chosen parishes.

Influence of ethnicity on land use and management diversity

Discussions with local communities revealed that different ethnic groups had different approaches to biodiversity and land management. There were two subgroups among the Banyankole, according to their occupation and origin. People of the Bahima are traditionally nomadic cattle-keepers, while those of the other group, the Bairu, are traditionally settled cultivators. The Bahima used to depend almost entirely on their cattle, and traded with the Bairu to get some carbohydrate and other foods into their diet. They utilized the drier plains and hills, especially in agro-ecological zone B (see Table 3.3) for cattle-grazing and temporary homes. They moved from place to place in search of pasture and water for livestock. The transhumance patterns were associated with climatic seasons.

In this way, wild plant diversity was managed through rough rotational grazing and was thus protected from degradation since the system reduced chances of overgrazing. As the livestock grazed and moved, they also spread manure over the grazing areas and this acted as fertilizer and dispersal mechanisms to maintain vascular wild biodiversity. The herdsmen also lit bushfires towards the end of the dry season, a traditional practice for pasture management. Useful pasture species were therefore encouraged to grow at the beginning of the rains, but plants that could not survive bushfires were eliminated. Cattle-keeping thus involves relatively low levels of agrobiodiversity.

As cattle-keepers, the Bahima were primarily interested in cattle numbers for prestige and for paying a dowry. However, new socio-cultural values and changes in attitudes and market forces have led the cattle-keepers to change focus from numbers to quality. They are now becoming interested in rearing fewer cattle that give higher milk yields or grow faster and have larger carcass weights. This is due to changing social obligations and a need for money. For example, getting formal education for children requires school fees. Buying medical services and acquiring better housing and consumer goods require income rather than prestigious numbers of cattle. The market also prefers tender meat from faster-growing cattle to the traditional way of butchering aged, lean animals. A combination of the changes in social values, attitudes, obligations, and market preferences has therefore influenced the cattle-keepers either to upgrade their livestock through cross-breeding or to acquire exotic breeds.

Management of the livestock and pastures has also changed. The increasing population pressure on land has forced many Bahima to abandon their nomadic lifestyles for a more sedentary existence. Meanwhile, the land available for communal grazing is decreasing as more is opened up for cultivation. Because of the population pressure, marginal areas (which were previously reserved for grazing) are also being opened up for cultivation. Even the corridors through which the cattle used to be moved to grazing and watering points are being cultivated. As a result of all these changes, the cattle-keepers are becoming increasingly restricted to ranching or paddock grazing and more recently zero-grazing. They have also been forced into crop production to become more self-reliant for food and also to generate more cash by selling crops, including bananas.

To obtain adequate water supply for their livestock in the drier environments they occupy, the cattle-keepers have had to construct valley dams or water tanks for rainwater collection and storage in their farms, ranches, or communal grazing areas. Small herds are watered from springs and wells using water troughs. Therefore, the Bahima have had to diversify both the management practices related to their animals and the land uses they undertake in order to obtain their livelihoods. This has some positive implications for agricultural biodiversity, as will be discussed later in the chapter.

In a similar manner, the traditional cultivators, the Bairu and Bakiga, have adopted cattle-keeping as an additional source of livelihood and for balancing their diets. The Bakiga, who never used to grow bananas, have also adopted banana cultivation for similar socio-economic reasons.

Looking at the land-use types of the region, however, besides perennial banana growing which is almost everywhere and expanding rapidly, influences of the dominant ethnic group are still evident. The major subsistence crop of the sedentary Nkore people was millet, but in the last 50 years millet has been supplanted in large areas by bananas. The Bakiga-dominated Mwizi area is intensively cultivated, with many annual crops and minimal livestock production.

On the other hand, local people reported that traditions determine how the land is managed, but because of integration in settlement patterns the different ethnic groups have influenced one another and their ways of doing things (including land management practices and eating habits). These in turn influence the general agrobiodiversity. More specifically, seedbed preparation methods have a direct effect on soil fertility and soil water status. Some practices such as clean-tilth and trash-burning have degrading effects on soil biological, chemical, and physical properties, thereby influencing biodiversity and sustainability of agricultural systems. Land use and management have, therefore, substantially changed in response to social, demographic, economic, and environmental pressures. Different ethnic groups have taken on a selection of practices of other ethnic groups, and the result is now a far more agriculturally diverse and complex landscape.

Demonstration site level

During community workshops, Bushwere farmers reported major changes in agrobiodiversity over the past decade. Table 3.4 summarizes reasons for those changes.

Bushwere parish was found to have six land-use stages with many crop combinations distributed on all landscape types, as indicated in Table 3.5. The land-use stages included the following.

• Perennial banana-based with 13 crop combinations on the four different landscape types with 46 field types.
• Coffee-based land-use stage covering a small proportion of land with only five crop combinations on different landscapes, giving a total of 19 field types.
• Woodlots with only two field types, namely eucalyptus and coniferous. While eucalyptus was planted mostly on backslopes with a few patches on hilltops, the coniferous trees (*Pinus patula*) are grown in ravines. One expert farmer, Rubaramira, explained that woodlots are mostly feasible for people with large pieces of land and commonly used to

Table 3.4 Production trends for Bushwere over the past 10 years

Type of biodiversity	Reasons for decreasing	Reasons for increasing
Bananas (especially *Musa sapienta*)	*Nyamwenga* disease (*Fusarium scoparium*) on *Musa paradisca* in the valleys	Increased extension services Increased cultivated land Commercialization of banana crop
Coffee (*Coffea canephora Robusta*)	Bacterial wilt Low prices offered by coffee buyers Market monopoly/few traders	Good banana intercrop
Annuals, especially Irish potato (*Solanum tuberlosum*), maize (*Zea mays*), and peas (*Pisum sativa*)		Population increase Increased market demands Introduction of *taungya* land-use stage in gazetted land Increased extension services Increased/improved feeder roads
Finger millet (*Eleusine caracata*)	Declining soil fertility Labour demanding Low market demand	
Indigenous varieties of potato, maize, and beans	Introduction of high-yielding varieties Commercial market preferences	
Indigenous trees (e.g. *Combretum molle*)	Land clearing for cultivation Firewood harvesting Bushfires	
Eucalyptus species		Fast-growing trees to meet high demand for firewood and poles Grows on poor and infertile soils
Livestock	Decreased grazing land with more competing land uses Few distant watering points Increased labour shortage as most children go to school	

Table 3.5 Dominant land-use stages and their landscape positions in PLEC demonstration sites of Uganda

Land-use stage	Landscape			
	Hilltop	Backslope	Ravine	Valley
1. Perennial crops				
Banana agroforestry	*	*	*	*
Ba/Co/Ca/Mz/Bn	*	*	–	–
Ba/SWC/grass strips	*	*	*	*
Ba/Co	*	*	*	*
Ba home garden	*	*	*	*
Ba/Mz/Ca/Bn	*	*	*	*
Ba/MHg/trenches/grass strips	*	*	*	–
Ba/SWC	*	*	*	–
Ba/Mz	*	*	*	*
Ba	*	*	*	*
Ba (old)/Mz/Bn	*	*	*	–
Ba (young)/Mz/Mi	*	*	*	*
Sugarcane	P	P	*	*
Eucalyptus woodlot	P	P	–	–
Ba/Bn	*	*	*	–
Co	*	–	*	
Co/Ba (with trenches)	*	*	*	*
Co/Ba/Bn/Mz	*	*	*	*
Co/Mz/Bn	*	*	*	*
Co/Mz/So	*	*	*	*
Total	**20**	**19**	**18**	**20**
2. Annual crops				
Mi	–	*	–	–
Mi/Mz/Ca	*	*	–	*
Mi/Mz/So	*	*	–	*
Mi/Ca	*	*	–	*
Mi/Mz	*	*	–	–
Bn/Ca	*	*	–	–
Bn/I.Po	*	*	–	–
Bn/Mz/Ba (young)	*	*	*	–
I.Po/Ca	*	*	–	–
I.Po/Mz	*	*	–	–
Peas	*	*	–	–
So/Mi/Co (young)	*	*	–	*
So/Mz	–	*	–	–
So/Mi	*	*	–	*
Bn/Mz	*	*	*	*
Bn/Mz/Ca	*	*	*	*
Cabbages	–	–	–	*
Ca/Mz	*	*	–	*
Groundnuts	*	*	–	–
Groundnuts/Bn	*	*	–	–
Groundnuts/Ca	*	*	–	–
Groundnuts/Mz	*	*	–	–
I.Po	*	*	–	–

Table 3.5 (cont.)

Land-use stage	Landscape			
	Hilltop	Backslope	Ravine	Valley
I.Po/Bn	*	*	–	–
I.Po/Bn/Mz	*	*	–	–
I.Po/Bn/Ca	*	*	–	–
I.Po/Mz	*	*	–	–
Sweet potato	*	*	–	–
Sweet potato/Ca	*	*	–	–
Fallow (old)	*	*	*	–
Fallow (young)	*	*	*	*
Field edges	*	*	*	*
Grazed farm fallows	*	*	–	–
Ungrazed farm fallows	–	–	*	–
Total	**30**	**32**	**7**	**11**
3. **Home gardens**				
Pineapples	*	P	–	–
Mixed orchards	*	*	–	–
Vegetables	*	*	–	*
Agroforests	*	*	–	–
Compounds	*	*	–	*
Edges	*	*	*	*
Total	**6**	**6**	**1**	**3**
4. **Natural grassland**				
Combretum wooded grassland	–	*	–	–
Cymbopogon + Combretum + Pteridium	*	*	–	–
Hyperrhenia + Loudensia	*	*	–	–
Loudensia + Cymbopogon grassland	*	*	–	–
Fenced grazing land	*	–	–	–
Papyrus	–	–	–	*
Total	**4**	**6**	**1**	**3**
5. **Natural bushland/woodland**				
Natural forest (woodland)	*	P	*	–
Natural bushland (patches)	*	*	*	*
Pteridium woodland	–	*	–	–
Total	**2**	**3**	**2**	**1**
6. **Gazetted forest**				
Natural forest (woodland)	–	–	*	*
Planted forest (cyprus, pine, eucalyptus)	*	*	*	*
Natural reserve (wood grassland)	*	*	–	–
Taungya system (annual crops + young trees)	*	*	–	–
Total	**3**	**3**	**2**	**2**

Note: * = present; P = in patches; – = not available; Ba = banana; Ca = cassava; Co = coffee; Bn = beans; Mz = maize; G = grass; Mi = millet; I.Po = Irish potatoes; So = sorghum; SWC = soil and water conservation.

protect land from trespassers and bushfires. The pines are planted in ravines because they are "temporary" and release land for cultivation at harvest. On the other hand, eucalyptus coppices can stay on land as long as one wants, i.e. "permanently". Therefore it is planted on poor uncultivable backslopes. It is important to note that woodlots are very few and small in size. Many trees in Bushwere are scattered on farmland. The scattered trees are in clusters of three to 10 trees, along boundaries, in compounds, and as windbreaks around homesteads.

- Annual crop-based, with at least 34 crop combinations occurring on different landscape types to make 80 field types. Annual crops are less common in valleys and ravines, except cabbage, maize, beans, and cassava, and in most cases are used as an opening crop during the establishment of a banana crop.
- Home gardens, with 16 field types. Home gardens are defined here as fields that are within a distance of 100 metres radius from the homestead compound. These did not occur in ravines since the latter, being waterways, are not habitable.
- The edge field types are normally marking field boundaries (in the case of ravines) and are also found around homesteads as on other landscapes.
- Natural grassland, with nine field types mostly on hilltops and shoulder, on very shallow lataretic surfaces, climax species, and overgrazed except in recently settled hills bordering gazetted land.
- Natural bushland, with eight field types. These are patches of remnants of natural vegetation, mostly on uncultivable very steep backslopes, dominated by *Combretum* and *Pteridum* spp. Those in ravines, valleys, and hilltops belong to people with large pieces of land (early settlers).
- Gazetted land-use stage with 10 different field types.

During the biodiversity inventory exercise at least 194 distinct field types were encountered. However, due to limited resources only 20 of the most common and four uncultivated (natural) field types were studied in detail, and all were rich in species. More than 210 taxa (species) were documented in the 24 field types. Table 3.6 shows 21 of the field types studied and the average number of species in each.

Banana-based field types tend to have less species diversity than other field types, because of the clean culture management in banana gardens. Bananas are sensitive to competition, so weeds are normally removed as soon as they appear and most farmers practise mulching. These practices limit the number and abundance of species in these fields. Nevertheless, the banana crop creates a micro-ecosystem that favours biological processes and can support growth of many animal and plant species. Farmers take advantage of this to grow several different crops (intercropping) and trees (agroforests), and thus the many crop combinations in Table 3.6.

Table 3.6 Average number of species by field types assessed in increasing order of species diversity

Field type	Landscape position	Average number of species
Banana/maize/beans	Hilltop	16.8
Banana/maize/beans	Backslope	20.8
Beans/maize	Hilltop	22.0
Irish potatoes/maize	Hilltop	25.0
Banana/beans/maize/coffee	Backslope	25.5
Banana	Valley	27.3
Peas	Backslope	27.3
Irish potatoes	Hilltop	27.5
Loudentia/Hyperrhenia	Hilltop	28.8
Sorghum/maize	Backslope	29.0
Maize/millet	Backslope	29.3
Peas	Hilltop	30.3
Banana/maize/beans	Valley	32.0
Beans/maize	Backslope	32.8
Maize/beans/cassava	Backslope	33.8
Cassava/beans/maize/Irish potatoes	Backslope	34.0
Irish potatoes/beans/maize	Hilltop	34.3
Irish potatoes/maize	Hilltop	34.5
Cymbopogon/Loudentia	Footslope	36.3
Combretum/Hyperrhenia/Cymbopogon	Footslope	50.5
Pteridium/Combretum/savanna	Footslope	51.0

Most banana fields are multistorey systems. As long as appropriate spacing is ensured, productivity is not affected by intercropping. This probably explains why most of the useful plant species are conserved in banana fields (Nkwiine, Tumuhairwe, and Zake 2000). Natural grasslands have the highest species diversity because normally these field types are not interfered with, apart from grazing and occasional bushfires. Biodiversity is highest on backslopes and lower on hilltops which are more intensively grazed (*Loudentia* grasslands). Following grasslands closely is the Irish potato/maize (backslope) field type. This is probably because seedbeds are prepared by soil mixing which encourages germination of weed seeds, hence high diversity.

Household level

Agricultural diversification supports household livelihoods and contributes about 62 per cent of total household income. The most important income-earning crops in Bushwere are bananas, Irish potatoes, common beans, maize, sorghum, and millet. Some 30 per cent of Bushwere farmers reported that they depend on both off-farm and on-farm activities. However, most household earnings derive from off-farm activities such as carpentry, handicrafts, and trading in farm produce, and most of these activities are themselves dependent on agrobiodiversity.

Utility

It was observed in all households of Bushwere that people have many uses for the different plant species in their lands. Almost all the livelihood of each household depends on using plant parts in one or more ways, starting with construction and roofing the dwelling houses and granaries, and making furniture, tools, and implements. Timber, poles, and posts of different tree species are suited to different parts of the construction. Specific grasses and fibres of bananas are also suited for thatching or in construction (instead of nails), or for making strong ropes to tether livestock. Similarly, kitchen utensils and appliances such as pestles and mortars, mats, and baskets rely on agrobiodiversity. Banana leaves are used to make covers for bottles, calabashes, and pots, while certain trees or shrubs are used to make walking sticks and large boat-shaped vats for brewing beer. The traditional wooden basins, sandals, and stools are still common in some homes. Carvings and handicraft work in schools and women's clubs are also prevalent.

With such diverse and highly valued household uses of agricultural products, all household members benefit from the conservation of biodiversity. Fortunately, farmers still have access to officially demarcated government land which still has large areas of natural woodlands and grasslands conserved, although access is limited to specified non-destructive parts and quantities. These areas, however, are gradually being turned into planted forests. This, coupled with population pressure, will soon leave no space for useful wild plants unless deliberate conservation is planned and strategically incorporated in the farming practices.

Factors influencing conservation on farmland

Several factors underpin the capacity of households to conserve agrobiodiversity on their farmland. The main incentives are direct food, cash, and cultural and other socio-economic benefits. Traditional healers have conserved medicinal plants on their farmland, while artisans in construction, granaries, beehives, or handicrafts also try to conserve suitable species where possible or move long distances to purchase from people who

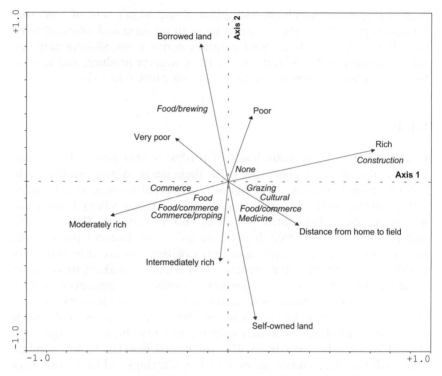

Figure 3.1 Species utility and socio-economic factors: canonical correspondence analysis
Note: The arrows indicate the socio-economic factors, while species utilities are indicated in italics.

have bushlands or woodlots. On the other hand, the main constraint is access to sufficient land. A canonical correspondence analysis (CCA) ordination relating biodiversity utilization to socio-economic categories of collaborating farmers and their access to land is presented in Figure 3.1.

Examining in greater detail the analysis in Figure 3.1, resource endowment appears to influence utilization of the biodiversity, and thus indirectly influence capacity to conserve it on the farm. The "poor" and "very poor" farmers generally cultivated borrowed land and thus were probably not able to conserve much biodiversity in their gardens – producing only crops utilized for food and brewing. Those in the "moderately" and "intermediately" rich categories, who are in the majority in the Bushwere community, cultivated their own land. They subsequently had much use for biodiversity on their farms. On the other hand, the "rich" farmers (who in most cases own a lot of land) were able to have woodlots and thus conserve some species for construction purposes.

However, they did not have as many utilities for different species as the "intermediate" category. Rich people in Bushwere were mostly traders, and did not invest their time in conservation management.

Distance of fields from the homestead also significantly influences species utility and thus conservation efforts. Conservation efforts of regularly used species are more prevalent around the homestead (Figure 3.2).

Other factors include:
- lack of knowledge of global, ecological, and other benefits of biodiversity beyond traditional values
- gender imbalances in the control of resources
- low levels of community mobilization for environmental and economic programmes; farmers operate as individual household units, which is not sustainable in these two aspects
- ineffective policies and regulations.

Models for sustainable agrobiodiversity conservation developed in Bushwere

From the preceeding observations and analysis, it is possible to derive "best case" examples of sustainable approaches for the integration of biodiversity conservation into agriculture. These approaches were developed by PLEC farmers and scientists on the Bushwere demonstration site, Mwizi (subcounty of Mabarara district). Demonstrations to other farmers and policy-makers have been centred on the following.
- Integrating stall-fed livestock into crop production systems, as demonstrated by expert farmer James Kaakare. His integrated plot in Figure 3.2 shows the interconnectedness of the biodiversity on different parts of the small-farm landscape utilized for efficient production purposes. Kaakare dug contour trenches across the banana plantation to conserve soil and water, which also improved his banana yields. Then he planted *setaria* grass to stabilize the soil bunds along the trenches and reduce the cost of cleaning trenches. Initially he used to cut the grasses to mulch, but soon learnt through interaction with PLEC scientists that they provide good fodder for livestock. He had also planted some cocoyams, vegetables, and fruit to improve his household's food nutritional status. With technical advice from PLEC as well as information gained from PLEC farmer-exchange visits, he was able to set up a stall-feeding unit for one milking cow. This was then an incentive to plant more fodder like Guatamala, callindra, Napier (elephant), and *kikuyu* grass within the banana plantation and homestead. He also feeds the cow with the male buds of the banana and fodder cut from other fields, where he was also encouraged to use grass bunds for soil and water

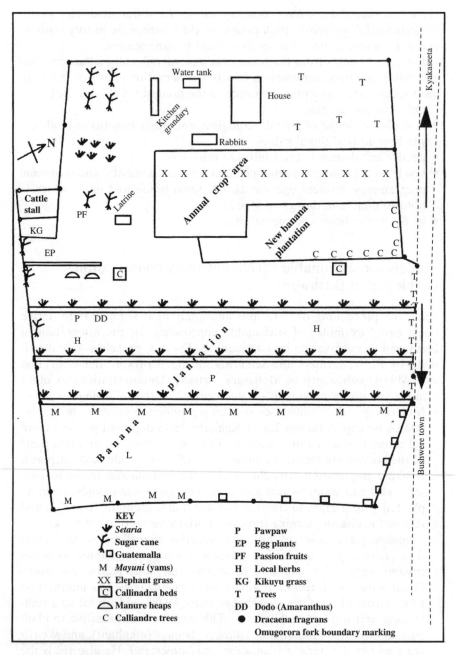

Figure 3.2 Sketch layout of James Kaakare's integrated biodiversity conserving system

Table 3.7 Plant diversity on Fred Tuhimbisibwe's farm

Functional grouping	Number of species
Medicinal	8
Food crops (banana varieties, grains, and vegetables)	16
Fruit trees	4
Fodder (legumes and grasses)	4
Coffee	1
Windbreaks (trees and two banana varieties)	4
Beehive and banana props	3
Boundary markers	2
Plant species in apiary:	
At edges	4
Within	over 10
Neighbouring fallow	over 50

Table 3.8 Fred Tuhimbisibwe's comments on his biodiversity conservation

- All flowers of these species provide nectar and pollen for bees.
- Several of these are multi-purpose.
- Sap from some species is used by bees to repair the combs and hives, some are fed to my goats and cow, and some give firewood and at the same time they fix nitrogen in soil.
- My management of biodiversity in this integrated system significantly improved through collaboration with PLEC scientists.
- I now have improved benefits and yields for food security and income and have a better environment in my farm.

conservation in annual crop fields. The cow provides manure for fertility improvement of the crops, and milk for his family. His income has increased significantly through sales of surplus milk and bananas. James Kaakare has trained many other farmers and has been instrumental in the Bushwere zero-grazing farmers' group.

- Integrating apiculture into mixed farms, as demonstrated by Fred Tuhimbisibwe. His farm has rich plant diversity, as shown in Table 3.7, much of which is encouraged by the need for nectar for his bees. The farmer's understanding of the agro-ecosystem is highlighted in Table 3.8. Retired lieutenant Fred Tuhimbisibwe was very innovative and had interplanted much of his fields to cater for both his household and his bees. Through PLEC scientists he developed his understanding of agronomic principles and selection of compatible species. As a result he improved his spatial arrangement and management, thus increasing the productivity of the food crop before he could be a convincing model to other farmers.

- Adding value to the sustainable use of biodiversity for farm structures and handicrafts. Examples include improved local granaries and modern maize cribs for better post-harvest handling and storage, demonstrated by Nevas Tugume, James Warugaba, and John Nyamwegyendaho; local and modern beehives demonstrated by D. Rubaramira; and bathroom sheds and fuel-saving stoves demonstrated by Joventa Kurigamba. PLEC provided technical guidance on proper ventilation for storage and also for potato chitting. The scientists also advocated use of metal rat guards, which reduced the amount of stored crops lost to vermin.
- Home gardens with different vegetables, fruits, herbs, and shrubs for ornamental and other uses, demonstrated by Joventa Kurigamba and Charles Byaruhanga.
- Integrating agronomic methods of soil and water conservation with banana production, as demonstrated by Frank Muhwezi.
- Development of common interest groups, including:
  - Bushwere Zero-grazing and Crop Integration Association (BUZECIA), with 24 members
  - Bushwere Nursery and Home Garden Farmers' Association (BUNUHOGAFA), with 29 members
  - Bushwere Development Group (BUDEG)
  - Mwizi PLEC Experimenting Farmers' Association (MPEFA), with 12 members.

These farmers' associations also serve as advocacy groups, by demonstrating to other farmers and policy-makers that household income, nutrition, community development, and environmental protection are all possible through integrating biodiversity conservation into agriculture. Farmers' associations monitor and motivate their members for continuity and achievement of objectives. One unique characteristic of these farmers' associations, and even the model technologies discussed above, is gender sensitivity and involvement of all household members in the approaches. Integrating conservation into agriculture and household work calendars is not feasible unless all members appreciate the values and share the benefits of biodiversity. This was the key that the PLEC-Uganda team discovered as they characterized the biodiversity potential of the smallholder farmers of south-western Uganda. Therefore all demonstration farmers had the support and assistance of their spouses and other household members. Realizing the benefits of gender balance and equitable sharing, the development-oriented farmers' associations (BUDEG, BUNUHOGAFA, and BUZECIA) all emphasize household membership and involvement in their constitutions. In fact the experimental farmers' group (MPEFA), though older, is not as strong as the other three because spouses and other household members were not involved during its formation. Even some field experiments failed due to lack of shared household responsibility among participating farmers.

• Income generation from plant nurseries demonstrated by BUDEG and BUNUHOGAFA, forestry demonstrated by D. Rubaramira, and herbal medicines demonstrated by the late Mzee Luka. Income-generating activities and projects are particularly advantageous in accessing credit facilities and community recognition. This promotes the programme's sustainability even beyond PLEC involvement.

Conclusion

Even under the pressures of economy, society, and population growth, agrobiodiversity flourishes in western Uganda. It is dynamic and ever-changing, supporting the objectives of both securing livelihoods and biodiversity conservation. With the focus of planning for the future, the following lessons have been learnt by the PLEC team in Uganda.

It was found that close collaboration between scientists and farmers on agrobiodiversity conservation increases the farmers' knowledge base on the ecological and economic (long-term and global) benefits of biodiversity conservation. It also stimulates their alertness and yearning for innovation in integrating compatible enterprises. The key is to stimulate their self-confidence and build on indigenous knowledge and skills.

Farmers with intermediate to moderate resource endowment have greater potential for integrating agrobiodiversity conservation into production through security of resource tenure and ability to manage and utilize the biodiversity resources. Efforts to promote sustainable agrobiodiversity conservation should target these categories of households initially – then others would follow. The rich farmers favour biodiversity that does not require regular management, for example trees (afforestation).

Conservation and sustainable use of agrobiodiversity depends on the direct benefits derived from undertaking these practices. Therefore, cost-benefit analyses should be included in technology development packages in order to screen good approaches for feasibility to promote adoption.

It was also established that farmer-led demonstrations, gender issues, and equitable sharing of benefits are fundamental to sustainable agrobiodiversity conservation efforts. Similarly, ethnic diversity and inter-marriages promote agrodiversity in general and agrobiodiversity in particular. Integrating biodiversity conservation and agriculture encourages the participation and training of all household members (including children) in sustainable resource management.

Associations and groups have a bigger voice to negotiate with policy-makers, markets, credit facilitators, and advisory service providers than individuals. Farmer groups/associations also have a wider impact and multiplier effect as well as being more concerned with community development. Therefore PLEC's work in agrobiodiversity conservation is more likely to be sustainable with the group and demonstration site approach than by working with individuals.

REFERENCES

Heywood, V. H. and S. D. Davies (eds). 1996. *Centres of Plant Diversity. A Guide and Strategy for their Conservation*. Gland: IUCN/WWF.

Nkwiine, C., J. K. Tumuhairwe, and J. Y. K. Zake. 1999. "Farmer selection of biophysical diversity for agricultural land uses in dissected plateaus of Mbarara, Uganda", in *Proceedings of the 17th Conference of Soil Science Society of East Africa*. Kampala: p. 304.

Statistics Department. 1992. *The 1991 Population and Housing Census Summaries*. Kampala: Republic of Uganda.

Zarin, D. J., H. Guo, and L. Enu-Kwesi. 2002. "Guidelines on the assessment of plant species diversity in agricultural lands", in H. Brookfield, C. Padoch, H. Parsons, and M. Stocking (eds) *Cultivating Biodiversity*. London: Intermediate Technology Publications, pp. 56–69.

4

Typical biodiversity in home gardens of Nduuri, Runyenjes, Embu, Kenya

John N. N. Kang'ara, Kajuju Kaburu, and Charles M. Rimui

Introduction

Nduuri is in the Runyenjes division of Embu district, lying on the slopes of Mount Kenya. It is situated between the forested Kirimiri hill to the north, which is the highest hill in the area, and the main road to Meru to the south. It is transected by three rivers (Thungu, Karii, and Kamiugu) which merge to form the River Ena, which eventually drains into the River Tana. Another main feature is Karue hill to the south, which, unlike Kirimiri, is covered with grass. The area is inhabited by the Embu tribe, most of whom are small-scale mixed farmers. The majority of land parcels have been demarcated and each parcel issued with a title deed. The amount of available land has been declining due to population pressure and the common Embu tradition of subdividing their land among their sons. At present the average land parcel size is about 1.5 ha per household.

Each household spares some land for the homestead, where residential, livestock, and storage structures are found. A number of plant species are also cultivated or deliberately allowed to grow wild on that land. Therefore, in order to understand the current state of play regarding agrobiodiversity and environment conservation, a survey was carried out in October 2000 to establish which plant species are grown (or being allowed to grow) in the homestead and how those species are being utilized.

57

Home visits were made by a research team consisting of a home economist, a livestock scientist/environmentalist, two field assistants, and a respected elder of the local community. The homestead area was determined using a tape measure and, together with a senior household member, the plant species were identified by type and niche, counted, and their utility recorded. Members of the research team, with the local community, then related the results to their significance for agrobiodiversity.

Main areas and groupings of plants

Homesteads

The portion of land spared for homestead use varied depending on a number of factors, including family size, the number and species of livestock kept, and total land area. Larger families kept larger areas for homestead than smaller families. Similarly, farmers with large farms assigned larger portions of land for homestead. Homestead size ranged from 0.04 to 0.3 ha, with an average of 0.144 ha. Since the homestead area often has the greatest agrobiodiversity in terms of species and varieties grown, there was found to be a significant positive relationship between family size and agrobiodiversity.

Species diversity

To show this agrobiodiversity, species distribution is presented in Table 4.1. Most of the species grown in the home garden (39 per cent) are food plants, the most popular of which is banana, which contributes 14.3 per cent of the 39 per cent devoted to food plants.

Fodder plants are also considered important, and make up 18.6 per

Table 4.1 Species distribution according to the utility in the homestead

Utility	%	Utility	%
Food	39.0	Hedge	3.0
Fodder	18.6	Fertility	2.4
Fuelwood	13.1	Art	2.0
Medicine	9.7	Support	1.8
Construction	7.9	Oil	0.6
Commercial	7.1	Mulching	0.4
Spices	4.2	Others	2.4
Ornament	4.0		

cent of the species grown in the homestead. Ornamental plants are grown for aesthetic rather than commercial reasons and are typically planted by young couples or around a newly constructed permanent house. Spices (the most popular being onion, chilli, and rosemary) are planted in the home garden for ease of accessibility during cooking. As arts and crafts are not popular within the Nduuri community there is no deliberate effort to plant suitable trees in the home garden. Most are there for other purposes, except for *muvuti*, an indigenous tree that appears only to have existence value to local people. Among the three plants associated with fertility (*Ficus sycamorus*, *Ficus thoningii* or *mugumo*, and *Tithonia diversifolia*), only *Ficus sycamorus* is planted deliberately for this purpose.

Species diversity, therefore, as the most obvious aspect of agrobiodiversity, varies considerably according to a wide variety of controlling factors. Wealth status, family size, local preferences, and recognition of the fertility-enhancing capability of plants all play key roles in determining the number of species planted and managed.

Hedges

Observation of the local landscape suggests that the greatest biodiversity exists in hedgerows, where the number of species in a small area is significantly greater than elsewhere. Sixty per cent of the homesteads visited had a perimeter live fence (hedge), sometimes reinforced with barbed wire. Within the homestead some houses were surrounded by small hedges (especially those used by the young sons) to separate them from other household structures. Contrary to initial speculation, however, the main line of hedges provides only 3 per cent of the plant species growing in the home garden. This statistic is deceptive because there are considerably more species associated with hedges and growing close to the hedge lines. Many of these associated species have local use value. The main species used for hedging included:

- *Leucaena leucocephala*
- *Lantana camara*
- *Caesalpinia volkensii*
- *Keiapple*
- *Kariaria* (*Euphorbia* spp.)
- *Cupressus* spp.

Choice of species for hedges was based on the ease of establishment, speed of growth, ability to prevent trespassers, appearance, and resilience. As Table 4.2 illustrates, other plant species often grow alongside or in conjunction with the hedges as their main niche. They are allowed to grow on their own or are deliberately grown for their other uses.

Table 4.2 Species growing alongside or in the hedge

Local name	Botanical name	Utility
Cypress	*Cupressus lusitanica*	Timber and fuel
Bougainvillea	*Bougainvillea formosa*	Ornamental
Mukwego	*Bridelia micrantha*	Fuel and construction
Calliandra	*Calliandra colothyrsus*	Fodder and fuel
Mururi	*Commiphora zimmermannii*	Fodder, yam support, and boundary maker
Mukinduri	*Croton megalocapus*	Fuel
Ithare	*Dracaena steudneri*	Fodder and ornamental
Mukima	*Gravellia robusta*	Construction and fuel
Leucaena	*Leucaena leucocephala*	Fodder
Passion fruit	*Passiflora edulis*	Food
Napier	*Pennisetum purpureum*	Fodder
Kirurite	*Tinothia diversifolia*	Fodder and medicinal
Mubiru	*Vangueria madagascarensis*	Food and fuel
Muhuru	*Vitex keniensis*	Construction and fuel
Mucimoro	*Lantana camara*	Fodder and fuel
Yam	*Dioscorea minutiflora*	Food
Eucalyptus	*Eucalyptus saligna*	Construction and fuel
Jacaranda	*Jacaranda misomifolia*	Construction, ornamental, and fuel
Muriria		Food and medicinal
Mutundu	*Neoboutonia macrocalyx*	Fodder, fuel, and medicinal
Velvet beans	*Mucuna pruriens*	Fodder and food
Pine	*Pinus patula*	Construction and fuel
Muterendu	*Teclea nobilis*	Construction and fuel
Munyenyenga	*Cyphostema maranguens*	Medicinal
Murenda	*Hibiscus trionum*	Construction and fuel
Muu	*Markhamia lutea*	Construction and fuel
Mungirima	*Ochna ovata*	Art and craft
Pine	*Pinus patula*	Fuel and construction
Sesbania	*Sesbania sesban*	Fodder
Sorghum	*Sorghum bicolor*	Food

Food plants

The majority of food crops grown in the home garden are planted, while others grow of their own accord. Fruit plants constitute 67 per cent of all cultivated plants, of which passion fruit is the most popular, accounting for 39 per cent of the fruit plants grown, followed by bananas (13.8 per cent), and mangoes (see Table 4.3). There are more than 10 different varieties of banana grown in Nduuri, of which *Kiganda*, *Miraru*, *Gacuru*, and *Kampala* are the most popular. Mango, banana, and passion fruit are dominant because they are commercialized and, as well as being used for household consumption, can provide a substantial income. In old home-

Table 4.3 Food plant diversity in Nduuri

Local name	Notes
Passion fruit	Commercialized
Apple	A few households
Avocado	
Banana	Very popular
Cape gooseberry	
Pigeon pea	
Cassava	Increasing commercialization due to poor coffee prices
Climbing bean	Newly introduced
Bean	
Yams	
Coffee	Not very common in home gardens
Cowpea	Popularly used as a vegetable
Guava	
Pumpkin	
Lemon	Declining use due to greening disease
Loquat	
Macadamia	Mainly in the old homestead
Amaranth	Wild or planted
Maize	More than seven varieties
Black nightshade (*managu*)	
Mangoes	
Masecondari	Fairly new to the area
Matomoko	
Mbumbu	Indigenous climbing perennial bean
Muriaria	Planted or wild indigenous vegetable, also medicinal
Muviru	Wild fruit
Mukaurivu	
Mulberry	
Mukwini	
Murenda	
Nduuma (arrowroot)	
Ngambura	Wild fruit, also medicinal
Onion	
Pawpaw	
Peas	
Pineapple	
Potatoes	
Sweet potatoes	Also used as fodder
Sorghum	Declining due to bird attack
Soya bean	Newly introduced
Spinach	
Sugarcane	
Sukuma wiki (kales)	Used as a vegetable
Sunflower	
Tree tomatoes	
Velvet bean	Used for brewing beverages and also as fodder

steads macadamia constitutes about 2 per cent of the total food plants, but almost all of it is commercialized. Meanwhile, arrowroot (commonly known as *nduuma*), sugarcane, and kales (*sukuma wiki*) occupy the wet part of the home garden. This is typically close to a waste-water outlet and/or in a furrow or trench along the house edge to utilize rainwater from the roof top. Such efforts make it possible to grow water-loving plants in the homestead instead of the usual riverside or wetlands niche. Other food plants, especially vegetables and spices, are planted in the home garden in order to be accessed easily during cooking. Agrobiodiversity amongst food plants is strongly affected by soil fertility and micro-topography of food-crop plots.

Fodder plants

Most households have livestock, which play a major role in recycling nutrients in the farm and the home garden. Although little fodder grass is grown, crop residues play a major role in livestock maintenance. Banana stems and leaves are important as a source of forage, followed by multi-purpose trees such as *Leucaena*, *Calliandra*, and mulberry (see Table 4.4). Being a popular fruit and staple food plant for the Embu commu-

Table 4.4 Fodder plants grown in Nduuri

Local name	Description and other uses
Avocado	Food (only local avocados – grafted are not used)
Bananas	Food
Calliandra	Mainly used as fodder
Leucaena	Mainly used as fodder
Comelina	Weed
Cassava	Food
Tithonia	Weed
Lantana	Weed
Masecodari	Food
Triumfetta (*mugico*)	Weed
Mukwego	Mainly for goats, a multi-purpose indigenous plant
Mugumo	Mainly for goats, a multi-purpose indigenous plant
Muhehe	Weed
Mulberry	Mainly used as fodder
Muria	Weed
Mururi	Supporter for yams
Mutundu	Mainly for goats, a multi-purpose indigenous plant
Napier grass	Mainly fodder
Sweet potato	Food
Sesbania	A multi-purpose fodder plant
Sugarcane	Tops used for fodder
Maize	Leaves and stover used for fodder
Mutei	Weed

Table 4.5 Medicinal plants found in home gardens

Local name	Description and other uses
Castor oil plant	Lubricant oil
Conge	Weed
Mukinduri	Fuel
Giatha	Weed
Muriria	Weed and vegetable
Kirurite	Weed and fertilizer
Kithunju	Purely medicinal use
Lemon	Food
Melia spp.	Construction and fuel
Miraa	Commercial stimulant drug
Muvuthi	Purely medicinal use
Mukandu	Weed
Munyenyenga	Weed
Mutei	Weed
Mutundu	Fuel and fodder
Mwinu	Weed
Mutongu	Weed
Mukambura	Fuel and food
Pawpaw	Food
Chillis	Spice
Pumpkin	Food
Mwiria	Construction

nity, bananas are planted by most households and their by-products are used for feeding stall-fed ruminants. Some indigenous trees with several uses such as *mukwego*, *lantana*, *mururi*, and *mutundu* are deliberately allowed to grow along the hedges, or are conserved in the homestead for feeding goats. Most multi-purpose fodder trees are grown as hedges. Although avocado is indicated as a fodder for goats and cattle, farmers reported that grafted avocado trees were highly toxic to the animals and hence were not used as feed. Weeds such as *Comelina benghalensis* and natural grasses play a major role in livestock feeding and are allowed to grow in the garden before being removed and used as livestock feed. A few households grow Napier grass in their home garden, but most forage is obtained away from the homestead.

Medicinal plants

Medicinal plants are well known for their significant contribution to overall agrobiodiversity. They consist of many species – some wild, some domesticated, and some intensively managed and used for other purposes. Although medicinal plants constitute 9.8 per cent of home garden agrobiodiversity, it was found that the majority are not grown solely for medicinal use (see Table 4.5). Only *muvuthi*, *ndawa ya mariria*, and

Table 4.6 Species used for fuelwood and construction

Local name	Description and uses
Cypress	Hedges, fuel, and construction
Eucalyptus	Fuel and construction
Grevillea	Fuel and construction
Macadamia	Commercial fruit and fuel
Muterendu	Fuel and construction
Muviru	Food and fuel
Muhuru	Fuel and construction
Croton	Shelter, medicinal, and fuel
Mukwego	Construction and fuel
Muu	Fuel and construction
Mwiria	Medicinal, construction, and fuel
Mucavavunduki	Fuel and construction
Pine	Fuel and construction
Avocado	Food and fuel
Cassia	Fuel
Coffee	Fuel and commercial
Mugumo	Fodder and fuel
Guava	Food and fuel
Loquat	Food and fuel
Jacaranda	Fuel and construction
Leucaena	Fodder and fuel
Melia spp.	Fuel
Lantana	Fodder and fuel
Mukuyu	Shelter and fuel
Muhuru	Construction and fuel
Mukurwe	Fuel
Mutundu	Fuel and fodder
Umbrella tree	Shelter and fuel

kithunju (aloe) are grown in the home garden specifically for the treatment of malaria, wounds, and chicken ailments. Some, like *mukandu*, *kirurite*, *mutei*, *giatha*, *munyenyenga*, *ndongu*, and *conge*, grow naturally as weeds but have recognized medicinal uses.

Wood for fuel and construction

Plants used for fuelwood and construction are summarized in Table 4.6. While all construction woods can also be used as fuelwood, not all fuelwood can be used for construction purposes. Fuelwood constitutes 13.1 per cent, while the construction woods make up only 7.9 per cent of the home garden biodiversity. About 15 species found in home gardens are used for construction. Most of them are commonly planted exotic trees. Many households neglected their cypress hedges after they were struck by an outbreak of an aphid-transmitted viral disease. Surviving hedges

Table 4.7 Commercial crops

Local name	Utility
Avocado	Food
Banana	Food
Castor oil seed	Oil
Coffee	Beverage
Cypress	Construction
Eucalyptus	Construction
Grevillea	Construction
Miraa	Stimulant
Macadamia	Food
Passion fruit	Food
Mangoes	Food
Soya bean	Food
Tobacco	Stimulant

subsequently grew to become trees and were then of a size to be sawn into timber. Among the planted trees, *Grevillea robusta* is the most popular because it grows fast and combines well with crops and coppices to yield fuel and fencing material for domestic use. Fuelwood was also provided by crop prunings from trees such as avocado, macadamia, and guava and loquat branches and stems of coffee.

Commercial plants

Commercial crops range from conventional cash crops like coffee to food crops such as bananas, as summarized in Table 4.7. *Miraa*, a commercial stimulant popular in the Horn of Africa, was introduced in the area as an alternative source of income in the light of declining coffee prices. Meanwhile, banana's successful commercialization is due to its proligacy, multi-purpose uses, fast growth, and ready market, especially for the varieties *Kampala*, *Kiganda*, and *Israel*. Eucalyptus is now popular for providing roofing material for permanent houses following the closure of forests to local sawmillers. One eucalyptus tree has reportedly fetched close to Ksh12,000 ($150) for timber.

Conclusion

Although home gardens are relatively small (average 0.144 ha), the diversity of species is remarkable, changing in response to production opportunities, population pressure, and competing uses. In terms of overall agrodiversity in the landscape, home gardens and associated areas (such as hedges and fuelwood plots) are the most significant for agrobiodiversity and management diversity.

The principal driving forces for agrobiodiversity in these home gardens are the different uses for plants. Several species have multiple uses and hence are most commonly found within the homestead. Food use is the most common utility, but is closely followed in terms of numbers of species documented by fodder, fuel, and medicinal uses. There is also increasing interest in plants with commercial value which can derive substantial income. The species' requirement of inputs – whether the plants require planting and subsequent tending or whether they occur naturally – is also a factor and influences its popularity.

Agricultural biodiversity is thriving in the home gardens of Nduuri, Kenya, and there is much evidence that its dynamism is supporting the livelihoods of local people against threats, such as coffee berry disease, which challenge individual species. These findings emphasize the importance of directing conservation efforts towards home gardens because of their key role in agrobiodiversity. In Part II of this book, a selection of individual practices and management techniques will highlight how this biodiversity is being managed to the benefit of both development and conservation. This will then demonstrate that not only is agrobiodiversity rich in home gardens, but management diversity is also, with a wide range of techniques and methods, most of them derived from local knowledge.

Part II

Components of agricultural biodiversity

5

Spatial and temporal characteristics of rainfall in Arumeru district, Arusha region, Tanzania

Robert M. L. Kingamkono and Fidelis B. S. Kaihura

Introduction

Chapter 2 introduced the various aspects of diversity encompassed in the use of the term "agrodiversity". One important background aspect is the natural diversity exerted by the biophysical environment, which both presents a challenge to land users and provides an opportunity for exploitation of the variety in order to build livelihoods. Especially in low-input farming systems, biophysical diversity controls the intrinsic quality of the natural resource base which is utilized for production. It contains the natural resilience of the biophysical environment, to be exploited by land users. It includes soil and climatic variability. It takes account of physical and biological processes, hydrology, microclimate, and also variability and variation in all these elements. Farmers select within this diversity and they often manipulate it quite substantially. In the case of rainfall, the subject of this chapter, farmers employ a variety of techniques to mitigate droughts and floods, as well as to make most efficient use of rainfall for productive purposes. Rainfall harvesting is a technique, for example, that has gained prominence in African drylands as a flexible coping strategy to deal with the vagaries of nature (Adams and Mortimore 1997). One of the principal ways in which farmers have learned to cope is through agrodiversity.

Rainfall is the aspect of the biophysical environment that most farmers mention when they talk about their difficulties in managing their farms and maintaining production. It is a common phenomenon for all PLEC demonstration sites, but especially for semi-arid and sub-humid zones, as found on the lower slopes of Mount Meru in northern Tanzania. The unreliability of rains and their variation from year to year, and within season, all give rise to regular complaints. The variability in both space and time of rains is an aspect of the diversity of the biophysical environment with which farmers have learnt to cope. Indeed, other chapters in this volume show that biophysical diversity is an aspect that actually promotes agricultural biodiversity, with the planting of various species and varieties designed to mitigate risk and secure food supplies in the eventuality of extreme variability of rainfall. Understanding this inter-relationship between biophysical diversity and agricultural biodiversity is one essential aspect of the process linkages in the "agrodiversity framework" presented in Chapter 2. Linking this understanding to farmers' perceptions and views on the role of rainfall variability in their farming practices is especially important in throwing light on both the constraints and the opportunities afforded by such variability and its role in maintaining on-farm biodiversity.

The study of climatic change through historical meteorological records has a long provenance in Africa (Hulme 1996). Here, an analysis has been conducted of the spatial and temporal trends of rainfall in Arumeru district, Arusha region, using relatively conventional techniques and long-term records. The specific objectives were as follows:

- to study the spatial and temporal seasonal rainfall trends within and between ecological zones of the district – high altitude/high rainfall, middle altitude/average rainfall, and low altitude/low rainfall zones
- to determine the onset and end of rains in the district
- to assess the length of the growing season in the district
- to examine the occurrence of dry spells within the growing season and identify any changes
- to investigate farmers' views on weather changes and how these affect farming practices and daily life.

Daily rainfall data from 13 stations were collected from the Department of Meteorology in Dar es Salaam. The stations, their location, and altitude are shown in Table 5.1. The table also gives the mean annual rainfall and the duration of rainfall records used in the analysis. The shortest rainfall record was for Tengeru coffee estate with 15 years (1961–62 to 1975–76) and the longest was for Olmotonyi with 66 years (1927–28 to 1992–93) of data. Most stations have missing data for individual rainfall events, and when this extended over several months the years under which they belong were coded as having no data. A participatory rural appraisal (PRA) was then carried out to gather farmers' opinion of the weather changes and collect their views on the extent to which such changes affected their daily life.

Table 5.1 Details of stations used in the study

Station	Location	Altitude (metres above sea level)	Mean seasonal rainfall (mm)	Duration of rainfall record analysed
Ngurdoto Crater	3°18'S 36°55'E	1,676	1,185	1962–63 to 1993–94
Narok forest station	3°20'S 36°40'E	1,829	1,985	1961–62 to 1991–92
Olmotonyi	3°18'S 36°39'E	1,609	915	1927–28 to 1992–93
TPRI	3°20'S 36°37'E	1,432	805	1954–55 to 1996–97
Selian coffee estate	3°21'S 36°36'E	1,402	921	1933–34 to 1996–97
Arusha airport	3°22'S 36°38'E	1,387	862	1959–60 to 1997–98
Arusha Maji	3°23'S 36°41'E	1,402	1,005	1965–66 to 1996–97
Themi	3°24'S 36°42'E	1,372	1,097	1935–36 to 1989–90
Tengeru coffee estate	3°22'S 36°48'E	1,219	1,183	1961–62 to 1975–76
USA Ltd	3°23'S 36°52'E	N/A	826	1973–74 to 1996–97
Dolly estate	3°25'S 36°52'E	1,067	745	1934–35 to 1987–88
KIA	3°25'S 37°04'E	891	525	1971–72 to 1997–98
Lucy sisal estate	3°33'S 36°49'E	900	426	1976–77 to 1995–96

Note: N/A = not available.

There are two distinct rainy seasons in the district: the short rains (*vuli*) and the long rains (*masika*). The *vuli* rains are unpredictable, and in some stations (especially those at lower altitudes) and some seasons they do not occur at all.

General trend of seasonal rainfall

The general trends in seasonal rainfall are summarized in Table 5.2. Three stations were grouped under the high-altitude zone (Olmotonyi, Ngudoto Crater, and Narok forest station). Stations in the middle-altitude zone included Tenguru, Themi, Arusha Maji, Arusha airport, Selian coffee estate, and TPRI. Finally, the low-altitude zone consisted of the Dolly estate, KIA, and the Lucy sisal estate.

The start, end, and length of rainy seasons

Definitions

The start of the rainy season was the first date when at least 10 mm of rain had fallen over seven consecutive days, with rain falling on at least

Table 5.2 General trend of seasonal rainfall in stations of different altitudes

Zone	Altitude range (metres above sea level)	Rainfall range (mm)	General trend	Other observations
High	1,609–1,825	915–1,985	Rainfall at Ngurdoto Crater constant over time; at Narok forest station rainfall fell after 1983; at Olmotonyi rainfall decreased after 1972	Rainfall increases with altitude
Medium	1,219–1,432	805–1,183	Seasonal rainfall fairly constant; no cyclic trends although at Tengeru rainfall decreased in the early 1970s	Rainfall higher in the east
Low	Below 1,100	426–745	No cyclic trends; rainfall at the Lucy sisal estate is decreasing	Rainfall increases with altitude

three of those days. The end of the rains was the first date on which 15 consecutive dry days had occurred. A day was considered dry if it had less than 1 mm of recorded rainfall (such an amount is often insignificant in terms of its contribution to crops as it is lost through evaporation before being available to the plant). Therefore a day was considered wet if there was 1 mm or more rain. The length of season was the duration between the start and end dates of the rains. By those definitions the *vuli* rainy season in Arumeru district was 1 October to 1 November while the season of *masika* rains was 15 January to 1 April.

Start and end of *vuli rains*

The start and end dates of the *vuli* rains for each station were determined for each year. Frequency distributions of these dates were determined and percentage points at 20, 50, and 80 per cent derived. As shown in Table 5.3, the high-altitude zone sees the *vuli* rains (according to the median start dates) starting in the first half of November on the eastern side of the district and advancing towards the western side. The median end dates of the *vuli* rains are in the second half of December and first week of January. The rain ends earlier on the eastern side than on the western side. There is some variability in start dates, with the coefficient of variation ranging between 28 per cent and 37.3 per cent. The coeffi-

Table 5.3 Probable start and end dates for the *vuli* season at given percentage probabilities

Station	Altitude zone	Start dates				End dates			
		20%	50%	80%	Coefficient of variation (%)	20%	50%	80%	Coefficient of variation (%)
Ngurdoto Crater	High	16 October	5 November	22 November	37.3	30 November	15 December	31 December	17.7
Narok forest station	High	18 October	10 November	19 November	28.0	11 December	7 January	25 January	19.5
Olmotonyi	High	30 October	13 November	28 November	30.4	8 December	23 December	22 January	31.6
TPRI	Medium	5 November	19 November	5 December	25.6	4 December	15 December	4 January	21.9
Selian coffee estate	Medium	4 November	20 November	14 December	39.1	21 November	7 December	29 December	20.1
Arusha airport	Medium	31 October	17 November	27 November	34.5	27 November	15 December	13 November	21.3
Arusha Maji	Medium	25 October	12 November	25 November	32.3	28 November	9 December	20 December	17.2
Themi	Medium	22 October	10 November	7 December	29.7	23 November	14 December	3 January	21.3
Tengeru coffee estate	Medium	22 October	11 November	6 December	56.3	21 November	1 December	17 December	21.6
USA Ltd	Medium	3 November	20 November	15 December	55.8	24 November	1 December	12 December	19.2
Dolly estate	Low	1 November	N/A	N/A	52.8	21 November	27 November	11 December	11.5
KIA	Low	8 November	25 November	N/A	52.2	21 November	30 November	8 December	15.8
Lucy sisal estate	Low	12 November	N/A	N/A	49.4	21 November	23 November	5 December	15.2

Note: N/A = not available.

cient of variation of end dates ranges from 17.7 per cent to 31.6 per cent.

In the middle-altitude zone, the median start dates indicate that *vuli* rains commence in the second half of November. The trend again (for the start and end of the rains) is from east to west. The median end dates of *vuli* rains in this zone are within the first half of December, ranging from 1 December at Tengeru and USA to 15 December at TPRI and Arusha airport. The coefficient of variation for both start and end dates also reveals only moderate variation (see Table 5.3).

The median start date for *vuli* rains in the low-altitude zone lies between late November and early December. In one out of five years the *vuli* rains are non-existent. The median end dates are within the second half of November. The start dates in this zone are highly variable, with coefficient of variation figures between 49.4 per cent and 52.8 per cent. End dates do not vary much, with coefficient of variation figures ranging from 11.5 per cent to 15.8 per cent.

The length of the vuli *season*

Similarly, the length of the *vuli* season was calculated for each year on every station (see Table 5.4). The *vuli* season is typically longest in the high-altitude zone (median length ranging from 40 days at Olmotonyi to 55 days at Narok forest station). Meanwhile, in the medium-altitude zone, the median length ranges from 17 days at USA to 40 days at Themi. In the low-altitude zone the median length ranges from zero (at the Dolly and Lucy estates) to just five days (at KIA), but in approximately one in five years there are no *vuli* rains at all. At the Lucy and Dolly estates there is only ever a 50 per cent chance of *vuli* rains. There was evidence of year-to-year variation in season length at every station, but there was generally more variation in the lower zones.

The start and end of the masika *season*

The start and end of rains for the *masika* season was determined in a similar manner as described for the *vuli* season. As summarized in Table 5.5, rainfall in the high-altitude zone starts early at the higher stations and ends early at lower stations. In the middle zone the *masika* rains start earlier at the western stations than the eastern stations. For example, the median start date at TPRI (4 February) is approximately a month earlier than at the more easterly Tengeru (1 March). The end of rains in this zone is almost the same, with all stations having their median end date within the last week of May and first week of June. The median start of

Table 5.4 Length of *vuli* and *masika* seasons for stations in Arumeru district at three levels of probability

Station	Length (*vuli*)				Length (*masika*)			
	80%	50%	20%	Coefficient of variation (%)	80%	50%	20%	Coefficient of variation (%)
Ngurdoto Crater	30	42	53	54.8	99	133	178	33.0
Narok forest station	35	55	90	55.4	115	127	166	28.4
Olmotonyi	22	40	73	84.0	85	106	125	24.9
TPRI	12	27	57	84.2	95	115	127	20.0
Selian coffee estate	0	21	49	91.3	86	105	124	22.5
Arusha airport	9	33	66	76.6	92	113	128	21.8
Arusha Maji	7	32	53	80.6	87	117	129	28.2
Themi	8	40	61	73.8	88	117	137	26.8
Tengeru coffee estate	8	23	55	85.0	71	93	116	43.1
USA Ltd	0	17	43	113.2	68	91	105	28.4
Dolly estate	0	0	24	150.5	37	71	105	54.0
KIA	0	5	37	122.0	56	96	109	32.7
Lucy sisal estate	0	0	19	154.7	25	48	65	57.0

Table 5.5 Dates and percentage probabilities for the start and finish of the *masika* season

Station	Start dates					End dates				
	20%	50%	80%	Coefficient of variation (%)		20%	50%	80%	Coefficient of variation (%)	
Ngurdoto Crater	9 February	23 February	9 March	11.7		5 June	9 July	16 August	12.2	
Narok forest station	25 January	3 February	19 February	13.4		28 May	14 June	3 August	11.0	
Olmotonyi	26 January	11 February	29 February	11.5		18 May	29 May	7 June	6.4	
TPRI	26 January	4 February	16 February	10.5		19 May	25 May	6 July	5.7	
Selian coffee estate	28 January	16 February	4 March	11.4		21 May	31 May	8 June	5.2	
Arusha airport	26 January	9 February	23 February	11.6		21 May	29 May	9 June	6.6	
Arusha Maji	30 January	14 February	18 March	13.1		25 May	6 June	10 July	8.5	
Themi	26 January	12 February	8 March	12.9		22 May	5 June	24 June	8.0	
Tengeru coffee estate	11 February	1 March	22 March	12.8		15 May	23 May	24 June	10.5	
USA Ltd	10 February	25 February	28 March	14.0		18 May	28 May	5 June	7.0	
Dolly estate	7 February	11 March	10 April	15.8		6 May	20 May	5 June	7.0	
KIA	5 February	19 March	4 April	15.7		25 May	2 June	17 June	5.7	
Lucy sisal estate	17 February	19 March	5 March	11.6		17 April	3 May	21 May	8.0	

masika rains for all stations in the low-altitude zone falls within the second half of March. The rains end earlier at the western stations in this zone. For example, the median end date of *masika* rains is 3 May at the Lucy estate, while at KIA it is 2 June.

There is much less variation in the start and end dates for *masika* than *vuli* rains. Across all stations, the coefficient of variation for start dates range from 10.5 per cent (at TPRI) to 15.8 per cent at the Dolly estate. Meanwhile, the coefficient of variation for end dates varies from 5.2 per cent at the Selian coffee estate to 12.2 per cent at Ngurdoto Crater.

Length of the masika season

The length of the *masika* season increases with altitude. In the high-altitude zone median length ranged from 106 days at Olmotonyi to 133 days at Ngurdoto Crater, the middle zone experienced rains ranging from 91 days (at USA) to 117 days (as Arusha Maji), while the rains in the lowest zone lasted for between 48 days (Lucy sisal estate) and 96 days (KIA). Variability in length of this season is not as high as for the *vuli* season. The coefficient of variation in lengths for the *masika* season varies from 20 per cent at TPRI to 57 per cent at the Lucy sisal estate.

Temporal trends in start and end dates of the masika season

The variation of start and end dates for the *masika* season was examined to identify any long-term changes. One station with long-term records of rainfall data was selected from each zone: Olmotonyi, the Selian coffee estate, and the Dolly estate, representing the high, medium, and low-altitude zones respectively. The trend of start and end of rains in these stations is shown in Figures 5.1 to 5.6.

In Figure 5.1, the positive trend of dates for the onset of rains at Olmotonyi is shown by the dotted line, indicating that rains have been starting late recently. The polynomial trend (represented by the bold line) is showing some cyclic trends, but these are progressively uprising, indicating a shift in the onset of rains. Whereas before 1960 rains had never started later than day 185 (3 March), Figure 5.1 illustrates numerous occasions after 1960 where rains started later. Figure 5.2 shows the trend of dates for the cessation of rains for this station. The slope of the linear trend is slightly negative, indicating that rains are now ending earlier than in the past. These trends of the rains starting late and ending earlier are resulting in a shorter growing season. The other two stations (the Selian and Dolly estates) show the same trend, the latter with increasing magnitudes.

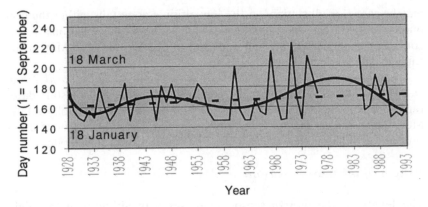

Figure 5.1 Trend of rainfall onset dates for Olmotonyi
Note: The dotted line and the bold curve are respectively the linear and polynomial trends.

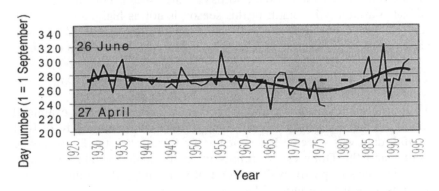

Figure 5.2 Trend of rainfall cessation dates for Olmotonyi
Note: The dotted line and the bold curve are respectively the linear and polynomial trends.

Occurrence of dry spells

Daily rainfall data were used to analyse the occurrence and risk of dry
spells. The same stations (Olmotonyi, the Selian coffee estate, and the
Dolly estate, representing the high, medium, and low-altitude zones)
were used for this analysis. It was performed over 30-day periods which
overlapped by five days. The maximum dry-spell lengths for each of the
30-day intervals were obtained and the proportion of years with dry
spells of at least 10 days was determined. Two types of analyses were

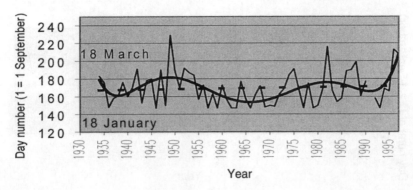

Figure 5.3 Trend of rainfall onset dates for Selian
Note: The dotted line and the bold curve are respectively the linear and polynomial trends.

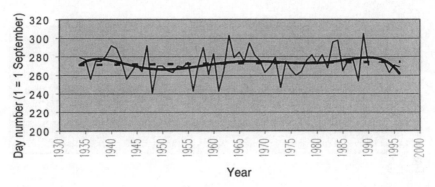

Figure 5.4 Trend of rainfall cessation dates for Selian
Note: The dotted line and the bold curve are respectively the linear and polynomial trends.

conducted – the trend of dry spells within a zone and the trend of dry
spells between zones.

Trend of dry spells within zones

The overlapping rainfall data for the three stations were divided into
three periods: 1934–35 to 1949–50; 1950–51 to 1969–70; and 1970–71 to
1987–88. The probability of the occurrence of a 10-day dry spell is shown
in Figures 5.7 to 5.9. For Olmotonyi, the probability of a 10-day dry spell
within the growing season was highest in the period 1971 to 1988 (see

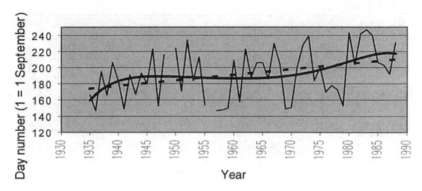

Figure 5.5 Trend of rainfall onset dates for Dolly estate
Note: The dotted line and the bold curve are respectively the linear and polynomial trends.

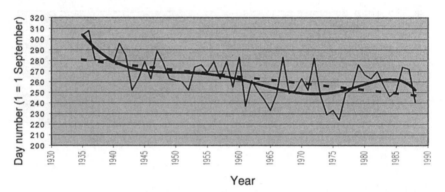

Figure 5.6 Trend of rainfall cessation dates for Dolly estate
Note: The dotted line and the bold curve are respectively the linear and polynomial trends.

Figure 5.7). For Selian the trend is different, as the same period saw the
lowest probabilities in the *vuli* season. The trend at the Dolly estate is
shown in Figure 5.9, in which a discernible increase in the occurrence of
dry spells in recent years is shown. While the probability of a 10-day dry
spell in April (between days 213 and 244) was less than 0.2 (sometimes
zero) in the period 1934–35 to 1949–50, in the period 1970–71 to 1987–88
it was between 0.5 to 0.8. Thus 80 per cent of the years since 1970 experienced a dry spell of at least 10 days in April.

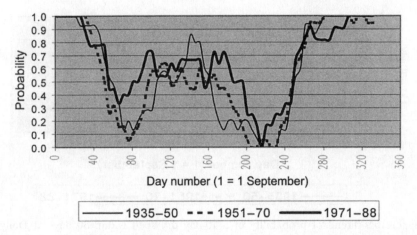

Figure 5.7 Trend of probability of a 10-day dry spell within 30 days at Olmotonyi

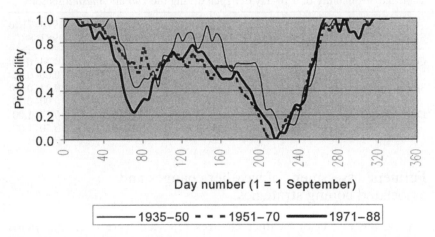

Figure 5.8 Trend of probability of a 10-day dry spell within 30 days at Selian

Trend of dry spells between zones

In the earlier period, the probability of 10-day dry spells at Olmotonyi ranged from 0.13 to 0.6 during the *vuli* season and between 0.0 and 0.6 during the *masika* season. These results are summarized in Table 5.6. It is evident that with decreasing altitude the probability of a dry spell increases.

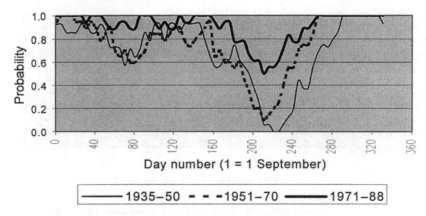

Figure 5.9 Trend of probability of a 10-day dry spell within 30 days at Dolly estate

Table 5.6 Probability of a 10-day dry spell during the *vuli* and *masika* seasons

Station	Season	1934–35 to 1949–50	1950–51 to 1969–70	1970–71 to 1987–88
Olmotonyi	*Vuli*	0.13–0.60	0.05–0.65	0.34–0.70
	Masika	0.00–0.60	0.00–0.70	0.00–0.75
Selian coffee estate	*Vuli*	0.42–0.90	0.50–0.70	0.22–0.78
	Masika	0.07–0.70	0.00–0.70	0.00–0.70
Dolly estate	*Vuli*	0.58–0.90	0.60–0.90	0.87–1.00
	Masika	0.00–0.80	0.10–0.90	0.50–1.00

Farmers' assessment of weather changes and associated coping strategies

Among the effects of weather changes reported were declining yields and food insecurity. Limited coping strategies were identified. Climate changes were considered more of a problem for crops than livestock. Crops such as *Dolichos lablab*, bulrush millet, and sorghum could no longer be grown. This was partly due to climatic variability, partly a lack of market, and partly the introduction of other crops such as maize, coffee, bananas, flowers, beans, vegetables, and Irish potatoes. Meanwhile, animals could be moved to other districts in Kiserian or sustained from pasture, crop residues, and house refuse.

Soil fertility management practices, including use of manure, incorporation of crop residues, planting trees in farm boundaries, and construction of stone lines for erosion control, continue to be carried out. However, fallowing practices disappeared with population pressure. Methods

Table 5.7 Farmer-assessed parameters and their level of importance in semi-arid and sub-humid Arumeru

Parameter	χ^2 significance level
Crop yield and food security	*
Coping strategies in livestock keeping	*
Change in crops commonly grown	***
Introduction of new crops	NS
Existence of traditional soil fertility management methods	*
Introduced soil fertility management methods	NS
Amount of water flowing from springs and rivers	*
Remaining sources of pastures and trees	*
Changes in onset of long rains	***
Changes in onset of short rains	*

introduced through development interventions included construction of contours (*fanya chini* and *fanya juu*), use of fertilizers, afforestation, and mixed/alley cropping with legumes and composting. Most of these methods experienced low adoption rates and were therefore not considered significant.

The flow of water from springs and small rivers was also indicated to have declined and in some cases ceased. This affected the length of time farmers could invest in irrigated vegetable production and off-season production of crops like green maize for roasting. Decrease in water flows was thought to be a result of deforestation, lack of conservation plans, extended drought periods, and agricultural development near water sources and/or in catchment areas. As a result, only hilly parts of the landscape remain as sources of pasture and trees as reserved/conserved forests. However, some farmers have introduced contour bunds (earth ridges on the contour) on their farms upon which pastures and trees can be planted. Both pastures and trees have improved on-farm biodiversity.

Conclusion

Rainfall variability appears to have increased significantly in all rainfall zones in recent years. This has affected the start and finish dates of both the short and long rains, and in lower rainfall areas it has even affected whether a viable *vuli* (short) season is available at all for production. Understanding the effect of these changes on how farmers cope with rainfall variability is a key challenge. One of the primary ways that farmers choose to mitigate the real problems of rainfall variability is to increase their on-farm biodiversity.

REFERENCES

Adams, W. M. and M. J. Mortimore. 1997. "Agricultural intensification and flexibility in the Nigerian Sahel", *Geographical Journal*, Vol. 163, No. 2, pp. 150–160.

Hulme, M. 1996. "Climate change within the period of meteorological records", in W. M. Adams, A. S. Goudie, and A. R. Orme (eds) *The Physical Geography of Africa*. Oxford: Oxford University Press, pp. 88–102.

6

The botanical knowledge of different age groups of the Nduuri farming community, Embu, Kenya

Ezekiah H. Ngoroi, John N. N. Kang'ara, and Charles M. Rimui

Introduction

The traditional method of transfer of knowledge and skills among the Embu ethnic group in Kenya was oral transmission. Information and skills were usually passed from father to son, grandfather to grandson, mother to daughter, grandmother to granddaughter, or from aunts to nieces. Information transfer could also cut across gender boundaries – from grandmother to grandson and so on. These transmission processes ensured that important information, such as knowledge of particular species and varieties and associated management practices, would not be lost from generation to generation.

Although the social structures that permitted this exchange still exist, there have been several changes which have tended to dilute their efficiency. Population pressure has led to a reduction in grazing land and therefore zero-grazing and tethering systems, which has reduced contact time between the old and the young. Traditionally, young boys looked after cattle and goats with their grandfathers and fathers all day. It was during these periods that older men would pass on knowledge and skills. With reduced contact time there is reduced transfer of knowledge and skills. Rural-to-urban migration has also separated the young (who have moved to urban areas) from the older generations (who have remained in the farmlands). This has also contributed to a reduced transfer of knowledge and skills.

The Nduuri community traditionally (and currently, although to varying degrees) relied on the biodiversity in their area for such needs as fodder, food, fuelwood, building materials, and medicine. Botanical knowledge underpinned the building of sustainable rural livelihoods by enabling the building of a substantial body of knowledge on how plants and domestic animals could be managed and protected.

This investigation was carried out in 2000 to investigate whether the Nduuri community was at risk of losing its traditional botanical knowledge. Through the Nduuri village sub-area headman, 10 elders were invited for a meeting. The elders were from five village sub-units, namely Gaturi, Karii, Kianjogu, Gicagi, and Kianjeru. The objective of the meeting was to define different age groups within the community and generate a list of the people in their village who belonged to those groups.

Formation of age groups was traditionally linked to the circumcision of young men. Those who were circumcised at the same time were considered to be of the same age group, in spite of actual age differences between them during the rite. The Nduuri elders could remember five age groups based on this traditional nomenclature. These were the Mugokoro, Njavani, Ngiciri, Ndururu, and Njenduru. However, as the rituals surrounding ciucumcision had changed, the elders noted that currently there existed no age groups as they understood them. The last of those age groups was the Njenduru of the 1930s. The elders therefore suggested using age sets covering periods of 20 years. Four age sets were proposed: 1920 to 1939 (over 61 years); 1940 to 1959 (41 to 60 years); 1960 to 1979 (21 to 40 years); and 1980 to 1990 (10 to 20 years).

Elders from the village sub-units were then requested to make lists of current residents according to their age set. Twenty names were then randomly selected from each set.

In individual interviews participants were asked to list names of plant species and their utilization. Participants were then grouped by age and asked to produce a consensus list. A further exercise was group identification of plant species in two marked quadrants – one in a recent fallow and another in a bush fallow. Finally, a discussion was held with the participants to analyse the outcome and suggest ways forward.

Knowledge of plant species

Different groups came up with various lists of plant species. The group consisting of the older-generation members (1920 to 1939) listed 180 species. The 1940 to 1959 group listed 195 species, while the 1960 to 1979 age set listed 153 species. Young people aged between 10 and 20 years listed 135 species.

The two oldest groups therefore knew the greatest number of species (180 and 195 respectively). One explanation of this could be that prior to the 1960s land parcels in Nduuri were not registered. There was therefore no intensive cultivation and many tree species. Now there are fewer tree species, the younger generation do not have knowledge of some of these now rare species. Additionally, livestock herders had access to free-range land and would get exposed to many of the species. The younger generation of the 1960s were born and brought up in individual land parcels, during a time when increased cultivation of cash crops (especially coffee) was leading to a reduction in wild vegetation.

Utilization of Nduuri's species

The participants identified a total of 68 ways in which the diverse species of Nduuri were utilized. These are summarized in Table 6.1.

Other uses not included in the table are poison, beverage making, musical instruments, jewellery, soil fertility improvement, storage structures, detergents, fermenting, milk preservation, composting, tobacco curing, food flavouring, and making dice for the *ajua* game. Others are witchcraft, bark extraction, bird trapping, dye, traditional ornaments, rain forecasting (by observing budding and flowering patterns), mattress stuffing, arrow colouring, fungicide, wedges for wood splitting, making utensils, and oil refining. The numbers of species for different uses mentioned by different age groups are summarized in Table 6.2.

The group that knew the most medicinal plants was the oldest (1920 to 1939, i.e. over 61 years), while the lowest number of species and their uses was found among the youngest group (1980 to 1990). The results suggest that the older generations valued and depended more on botanicals for medicine than later generations who were brought up during rapid expansion of medical facilities in the country. Another explanation is that older people have lived longer and have accumulated more knowledge.

The first explanation, however, is the more plausible. The 1960–1979 and 1980–1990 age groups were born and brought up during the expansion of education at and after independence. This, alongside changes in livestock management from free-range grazing to tethering and zero-grazing brought about by land demarcation in the early 1960s, led to changes in the role of many young people from tending cattle to attending school. These age groups were therefore not exposed to as many fodder species as their older relatives. Young people also no longer accompanied older family members on cattle-grazing trips and this loss of contact could have reduced knowledge transfer.

Beekeeping has traditionally been an occupation for the older genera-

Table 6.1 Utilization of Nduuri's plant species

Utilization	Notes
Medicinal	Various forms of treatment, whether through oral ingestion, massaging with hot leaves, antidotes, etc.
Fuelwood	Charcoal or firewood
Building	Used as poles or rafters
Food	
Timber	Species that can be split into timber either for construction or for making furniture
Fodder	Whether for direct grazing or "cut and carry"
Demarcation	Species used to mark boundaries; their basic characteristic was for permanency so they could survive through adverse weather conditions
Cleaning utensils and "steel wool"	Species with abrasive leaves used for scouring pots and pans
Yam support	Species that do not offer competition to the yam for nutrients
Pushcart wheel	Species that are light and will not easily crack
Banana ripening	Species whose leaves are placed in the ripening container and produce the bright colour of ripe bananas
Banana support	Tall varieties of bananas require support, especially by use of forked branches
Mole trapping	The mole is a common underground rodent pest of various crops; a mole trap consists of a small barrel made from stems of particular trees and two strings, one acting as the loop for strangling the mole
Oil	Usually for cosmetic purposes
Thatching	This is a rare occurrence currently as most households in Nduuri now have iron sheets for roofing
Fencing	Live fence or as posts
Tuck pins	Used for sticking photographs on the mud huts
Toothbrush	Species whose branches or roots are chewed to make a toothbrush
Tool handles	Hoes and axes that are frequently used require special hardwood handles to last long
Brooms	Made from shrubs with strong thin branches
Pesticide	By use of smoke or concoctions made from certain parts of the plant
Stimulant	Mild drugs that are legally accepted
Bee attractant	Used to prepare beehives in order to "invite" bees to colonize the hive
Beehive making	Species that are suitable for carving and whose timber will stand adverse weather conditions
Beehive support	Species that have good branching for safe beehive placement, conducive to bee colonization and honey harvesting
Latex extraction	For making waxes that were moulded into traditional spouts attached to beer gourds; sticky latex was also used for bird trapping by young boys

Table 6.1 (cont.)

Utilization	Notes
Terrace construction	For terrace stabilization
Fibre	Used for basketry, and as a substitute for nails for fastening rafters in construction
Walking sticks	Were a common feature among older people
Coffee shade	Trees that offer little competition to coffee but which give requisite shade to prevent "burning" of coffee plant leaves
Traditional tray	Basketry
Sandpaper	Leaves of such species have an abrasive texture for working on wooden furniture and utensils
Pestle and mortar	Strong timber with no cracks or weaknesses
Making weapons	Light but strong
Perfume	Aromatic species
Tissue paper/ handkerchiefs	Species with soft leaves which are mild to touch
Traditional candle	Species that burn slowly with more light than heat
Mats	Used for bedding
Gourd cleaning	The gourd is used for keeping a traditional beverage made of cereal flour, e.g. millet; special leaves are used to clean the gourd in order not to introduce off flavours in the porridge

Table 6.2 Use categories identified by different age groups

Age set	Total species	Me	Fw	Fo	Fa	Tr	Th	Hc	Pe	Bs	Fi	Ri
1920–1939	180	65	73	39	68	5	6	8	2	8	9	4
1940–1950	195	53	90	56	69	3	5	4	2	4	7	4
1960–1970	153	46	62	33	62	1	2	4	1	3	6	3
1980–1990	135	24	43	34	30	2	1	2	1	0	3	2

Note: Me = medicinal; Fw = fuelwood; Fo = fodder; Fa = fallow; Tr = trapping moles; Th = thatching; Hc = handcrafts; Pe = pesticides; Bs = banana support; Fi = fibre; Ri = ripening bananas.

tion. Honey was used for brewing, and also for matrimonial and land transactions conducted by older people. The older generation therefore appeared more knowledgeable in tree species useful as beehive supports (Table 6.3).

The 1960s ushered in a period of prosperity for many rural communities following the introduction of cash-crop farming and an expansion of education and job opportunities. Reduced numbers of species used for thatching coupled with increasing incomes meant that many people turned to using corrugated iron sheets to roof their homes. This could

Table 6.3 Other species and their uses

	Age set			
Use	1920–1939	1940–1959	1960–1979	1980–1990
To ripen bananas	*Mukinduri* *Mukorwe* *Muogoya* *Mutundu*	*Mucakaranda* *Mukinduri* *Mukorwe* *Muthithia*	*Mucakaranda* *Mukorwe* *Muogoya*	*Mucakaranda* *Mukorwe*
To trap moles	*Mubirairu* *Mukinyi* *Mukombokombo* *Murindangurwe* *Mururi*	*Mubiro* *Mukinyi* *Mururi*	*Mukinyi*	*Mukinyi* *Mwenjenje*
For thatching roofs	*Ikobo* *Mukobo* *Muthanje* *Nthagita* *Rukiriri* *Ruthiru*	*Ikobo* *Mukobo* *Muthanje* *Nthagita* *Ruthiru*	*Muthanje* *Ruthiru*	*Muthanje*
To make tool handles	*Mukinyi* *Murembu* *Murenda* *Muruva* *Muthigiriri* *Mutoo* *Muu* *Muvangua*	*Mucakaranda* *Murembu* *Murenda* *Mutoo*	*Mukinyi* *Murembu* *Muu* *Muvangua*	*Mukinyi* *Murenda*
As a pesticide	*Muvangi* *Muvuri*	*Kirurite* *Muvangi*	*Muvangi*	*Muvangi*
As beehive support	*Mucakuthe* *Mucuca wa Gitene* *Mugaa* *Mukarara* *Mukuu* *Mukuu* *Mumbu* *Mutanda*	*Mucakuthe* *Muganjuki* *Mugumo* *Mukuu*	*Mucakuthe* *Muthandam- bariki* *Muvuti*	
For fibre	*Mucico* *Mugaa* *Mukeu* *Mukonge* *Mukutha* *Murenda* *Muthanduku* *Mutoo* *Mwondoe*	*Mucico* *Mugere* *Mukeu* *Mukonge* *Murenda* *Murindangurwe* *Mwondoe*	*Mucico* *Mugaa* *Mugere* *Mukeu* *Mukonge* *Muthanduku*	*Mucico* *Mukeu* *Mukonge*

explain why the groups of the 1960s and 1990s appear to know fewer thatching species than the older generation (Table 6.3).

Identification of species in a recent fallow and bush fallow showed that the three oldest age groups were more knowledgeable than the youngest. This again could be because the younger group is little involved in farming activities (such as weeding and fodder gathering), as most of their time is spent in school. They are therefore less familiar with weeds and other species common in fallow land. The younger members of the community have also named some species for which the older generations did not have names.

Participants suggested ways of remedying the lack of knowledge among younger generations. One of the suggestions was to start herbariums in schools so that children could get acquainted with various species. This was considered particularly important for threatened species.

Conclusion

This study indicates that there are definite botanical knowledge gaps between age groups in Nduuri. Older generations are familiar with more species and their utilization, reflecting the changes in young people's role from agricultural tasks to schooling. However, further analysis into gender gaps of knowledge would be necessary before an action plan to address this apparent adverse trend can be formulated.

7

Diversity of vegetables and fruits and their utilization among the Nduuri community of Embu, Kenya

Kajuju Kaburu, John N. N. Kang'ara, Ezekiah H. Ngoroi, Seth Amboga, and Kaburu M'Ribu

Introduction

Nduuri is a catchment area on the eastern slopes of Mount Kenya where the PLEC project has one of its demonstration sites. The area has a bimodal pattern of rainfall that comes in March to May (long rains) and October to December (short rains). Farming in this area is small scale (typically 0.1 ha to 1 ha), with farming households relying significantly on cultivating fruit, nuts, perennial cash crops, and vegetables as well as their cereal staples. Maize and beans are the main food crops, while coffee, tea, and macadamia nuts are popular cash crops. Horticultural crops are grown as secondary crops. Coffee was the predominant cash crop, but it is now declining because of poor market conditions which have led to low financial returns for the berries. Some farmers have uprooted their coffee, while others continue to grow it in the hope that prices will improve. Meanwhile, new crops such as soyabeans and climbing beans have been recently introduced. Fertilizer use is limited. Instead, farmers use manure from livestock (predominantly dairy cattle and poultry) which is typically kept under zero-grazing, with fodder provided by the cut-and-carry system.

The objective of this survey was to carry out an inventory of vegetables and fruits, and determine how these crops are utilized and the behaviour of the farmers during times of gluts and shortages of these crops. A total of 41 farmers were selected using a combination of stratified and systematic sampling procedures. The area was divided into seven administration units and farmers from each unit were selected systematically for the interview. The number of farmers selected from each unit was relative to the size of its population, and ranged from five to nine.

A multidisciplinary team of research officers from the Kenya Agricultural Research Institute (KARI) and extension agents from the Ministry of Agriculture carried out interviews using a structured questionnaire. Three groups, each consisting of two research scientists and one extension agent, were formed and used as enumerators who assisted farmers in filling out the questionnaire. The groups were also required to confirm the farmers' information by observing farm activities. The questionnaire was designed to capture information on farming in general, and vegetable and fruit production in particular.

One day was devoted to initial briefing and testing of the questionnaire, after which the team was split into the respective groups. The survey was then carried out over a period of six days. On the final day of the survey an open forum was held between all informants, other interested farmers, and enumerators for further clarification of information. Data collected were then analysed using SPSS (Statistical Package for the Social Sciences) software.

Diversity of crops

Farmers grew both exotic and local types of vegetables (Table 7.1). Every home had a plot of cowpeas that was used both as a vegetable and a grain legume, while the most popular vegetables were found in over 50 per cent of farms. Crops were scattered in the farms on plots estimated to be around 20 square metres each. Some vegetables were planted in a few rows where grain crops were grown.

The most frequent local vegetables were cowpeas, amaranthus, pumpkins, tomatoes, and black nightshade, which were grown by more than half of the farmers. Local varieties of these vegetables as well as introduced ones were noted. These vegetables are commonly used in various parts of Kenya (M'Ribu, Neel, and Fretz 1993; Chweya 1997). A number of other indigenous vegetables were noted. They included *mariaria, karimi ka nthia, muka-urivu, makangati, magerema, maviu,* and *matanga,* but they were not properly identified or characterized. Many wild and semi-wild species are used as vegetables (Yongneng and Aigou 1999).

Other than some cowpea and pumpkin varieties, local vegetables were not deliberately planted but grew naturally and were tended. However,

Table 7.1 Fruit and vegetables grown

Vegetables	No. of farmers (n = 41)	Fruit	No. of farmers (n = 41)
Cowpea	41	Pawpaw	37
Amaranthus	39	Avocado	37
Kale	36	Banana	30
Pumpkin	34	Mango	29
Black nightshade	30	Passion fruit	27
Tomato	23	Guava	22
Spinach	12	Loquat	16
Mariaria	9	Pineapple	10
Karimi ka nthia	8	Mulberry	7
Chilli	7	Tree tomato	4
Makangati	6	*Macuca ma ngunga*	1
Russian comfrey	6	*Mubiru*	
Onion	6	*Mburu*	
Cabbage	5	*Ndoroma*	
Carrot	4		
Muka-urivu	3		
Magerema	1		
Matanga	1		
Maviu	1		

due to prolonged drought and continuous cultivation, most of the indigenous vegetables had disappeared or were in imminent danger of disappearing. Indigenous crops, although nutritious, are not popular with younger generations. Some were associated with poverty, while others were considered bitter or perceived as primitive. There was therefore a tendency towards exotic vegetables. This has also been observed in other parts of the country (Chweya 1997).

The most widely grown fruits were pawpaw, avocado, mango, banana, passion fruit, and guava (see Table 7.1). These fruits are popular in many parts of the country (M'Ribu, Neel, and Fretz 1993). Many farmers considered passion fruit, loquat, and guava as indigenous species, since they are long established in the area. The common indigenous fruits included *macuca ma ngunga*, *mubiru*, *mburu*, and *ndoroma*. These were not cultivated by farmers but were observed as individual plants scattered on the farms and in the forest areas.

As summarized in Table 7.2, farmers grew up to 14 species of vegetables and eight fruit species on their farms. Most grew four to six types of vegetable and five to seven species of fruit. All farmers grew more than three species, and most grew more species of vegetable than fruit.

Table 7.2 Diversity of vegetables and fruit grown per farm

No. of species per farm	No. of farmers in the category	
	Vegetables	Fruit
1	0	0
2	0	0
3	3	6
4	10	5
5	7	8
6	5	10
7	2	8
8	4	4
9	2	–
10	2	–
11	1	–
12	3	–
13	1	–
14	1	–

Table 7.3 Farmers' ranking of crops (crops appearing in the top three categories)

Vegetables	No. of times the crop was ranked	Fruits	No. of times the crop was ranked
Amaranthus	36	Pawpaw	37
Cowpea	33	Banana	29
Kale	29	Avocado	28
Pumpkin	25	Mango	18
Black nightshade	17	Passion fruit	6
Tomato	11	Tree tomato	2
Spinach	7	Pineapple	1
Russian comfrey	3	Mulberry	1
Cabbage	2	Guava	1
Chillies	1		
Mariaria	1		

Crop preference

Farmers ranked vegetables and fruit according to their production, preference, and utilization (see Table 7.3). Amaranthus, cowpeas, kale, and pumpkin were the most favoured vegetables, while pawpaw, banana, avocado, and mango were the most highly ranked fruits.

Table 7.4 Seasonal availability of vegetables in the farmers' fields

Season	Availability	
	Adequate (%)	Inadequate (%)
March to May	50.4	49.6
June to August	37.5	62.5
September to November	30.8	69.2
December to February	27.0	73.0

Table 7.5 Seasonal availability of fruits

Season	Availability	
	Adequate (%)	Inadequate (%)
March to May	29.2	70.8
June to August	31.8	68.2
September to November	18.4	81.6
December to February	17.4	82.6

Seasonal availability of fruits and vegetables

Vegetables were relatively available between March and May (see Table 7.4). Over half of the farmers (59 per cent) sell any excess vegetables, while others (27 per cent) consume more or share with friends and 12 per cent feed them to animals. Only 2 per cent preserve the excess vegetables for use during times of shortage.

As Table 7.5 demonstrates, there was no season when the majority of farmers felt that fruit production was adequate. This was confirmed by the small number of fruit trees observed on farms. The period between September and February had very low levels. Mangoes were available from late February to May, while avocado and pawpaw were available during the period between May and August.

When vegetables are scarce or absent, about 61 per cent of farmers buy extra from the market while the remaining 39 per cent do without vegetables during such periods. The most commonly bought vegetables are as shown in Table 7.6. Despite the scarcity of these products throughout the year, only a few fruits were bought for home consumption.

Utilization of fruits and vegetables

Although fruit production was low, farmers sold half of the fruit while the other half was consumed at home. Most commonly sold fruits were

Table 7.6 Percentage of farmers who buy vegetables and fruits

Crops bought	Farmers who buy (%)
Vegetables	
Kale	58.5
Cabbage	51.2
Cowpea	29.3
Tomato	7.3
Carrot	2.4
Fruits	
Passion fruit	14.6
Mango	12.2
Avocado	9.8
Pawpaw	4.9
Pineapple	2.4

pawpaw, avocado, and bananas. Fruits utilized at home were either eaten as they are, or together with other foods, or processed into juice, jam, and other products.

Most of the vegetables were used at home (75 per cent). Almost every home used vegetables for *ugali* and *githeri*. Pumpkin and cowpea leaves were used mainly for eating with mashed potatoes and/or bananas. Tomatoes were used for stewing with vegetables and meats. They could also be eaten as salad or used for making tomato sauce.

Sources of seed

For both vegetables and fruit, farmers either produced their own seed or purchased it. Local vegetable seeds were mostly produced on the farm, while the exotic ones were usually bought from shops. Farmers generally produced their own seed for local crops whose production potential is known (Chweya 1997). Many also traded in seed with their neighbours.

Pests and diseases affecting the crops

Diseases affecting vegetables included bacterial wilt, blight, mildew, and root rot. Being primarily subsistence crops, little effort was made to control disease in vegetables. Farmers did not pay much attention to crops that had no commercial value. However, some farmers used indigenous technical knowledge (ITK) and chemical methods to reduce these diseases.

Table 7.7 Percentage of farmers taking measures to control pests and disease in fruit and vegetables

	Pests		Diseases	
Measure	Vegetables	Fruits	Vegetables	Fruits
Chemical control	61	5	22	0
Indigenous technical knowledge (ITK)	19	2	15	0
None	20	93	63	100

Citrus fruits were affected by greening disease, while some types were being killed by mildews. The most common pests affecting vegetables and fruit were aphids, beetles, bollworms, and caterpillars. In vegetables they were controlled with chemicals and ITK, while little or no attempt was made to control such pests on fruits (see Table 7.7).

Conclusion

Nduuri's farmers grow a diversity of vegetables and fruit on their farms. Although these crops were widely grown in this area, they were considered as minor crops and production levels were low. Indeed, these crops were normally weeded last. There was a general shortage of products at most times of the year and thus low consumption. The survey showed that exotic vegetables were replacing traditional ones in the diet. There were many traditional wild and semi-wild species that have potential economic value as fruit, vegetables, or medicinal plants. Some of these species were available as early as the 1960s but are now considered as weeds and are in danger of becoming extinct. There seems to be a place for some of the indigenous vegetables, considering the percentage of farmers who had them on their farms.

Two main conclusions arise from this study of fruit and vegetables grown by the Nduuri community of Kenya. First, considering the nutritional significance of fruit and vegetables, it is incumbent on development agencies to pay greater attention to increasing total quantities of both fruits and vegetables. Strategies to increase supply of seedlings and seeds, as well as increasing the local base of knowledge on growing and managing these species, are indicated. Secondly, with regard to global objectives of protecting biodiversity, the farmers of Nduuri are indeed keeping a diversity of species, but the number of them is dwindling. Measures to preserve both species and knowledge about them could be beneficial. Growing these vegetables and fruits should therefore be encouraged as a way of conserving agrobiodiversity. Further work should be done on the specific varieties that are grown in order to retain those which have good potential for further breeding, production, and commercialization, as well as protection.

REFERENCES

Chweya, J. A. 1997. "Production and utilisation of traditional vegetables in Kenya", in *Progress and Prospect in Kenya's Horticultural Development Towards the Year 2000 and Beyond*, Workshop Proceedings. Nairobi: Kenya Agricultural Research Institute, pp. 33–39.

M'Ribu, K., P. L. Neel, and T. A. Fretz. 1993. "Horticulture in Kenya", *HortScience*, Vol. 28, pp. 779–781.

Yongneng, F. and C. Aigou. 1999. "Diversity of wild vegetables in Baka", *PLEC News and Views*, No. 12, pp. 17–19.

8

The effect of fig trees (*Ficus sycomorus*) on soil quality and coffee yield: An investigation into a traditional conservation practice in Embu district, Kenya

Ezekiah H. Ngoroi, Barrack Okoba, John N. N. Kang'ara, Seth Amboga, and Charles M. Rimui

Introduction

The Nduuri PLEC demonstration site is located in Kagaari, Embu district. It is within the agro-ecological zone classified as Upper Midlands 2 (UM) by Jaetzold and Schmidt (1983). The area receives about 1,250 mm of rainfall per annum and its soils are classified as humic nitosols.

In a participatory rural appraisal (PRA) conducted in 1998 the farmers indicated soil fertility as a major constraint to crop production in the area. The situation had been made worse by low coffee returns, as a result of which farmers were unable to purchase inorganic fertilizers to replenish fertility. Instead, they had devised several methods of replenishing soil fertility: namely, use of sweet potatoes in rotation, application of farmyard manure (FYM), retention *in situ* of crop residues, and planting fig trees (*Ficus sycomorus*) in coffee farms. Farmers claim the fig tree modifies the microclimate and improves soil fertility. Apart from providing food for livestock, birds, and wild animals (who eat the fruit), other uses include fuelwood, fruit, beehives, medicine (milky juice), mulch, ornamental, dune fixation, and traditional ceremonies (ICRAF 1992). Farmers reported that the idea of planting fig trees came from visits to the Thika area, a major coffee-growing area 100 km from Nduuri. By allocating trees to the coffee-growing areas, farmers would at the same time be conserving the species and adding to the biodiversity of the landscape.

This study was conducted to investigate whether the fig tree does fulfil the expectations outlined earlier and its benefits can be independently verified. The objective was therefore to assess the effect of the fig tree on soil quality and coffee yield. During a reconnaissance survey of Nduuri village, three farmers who were using the fig tree as a shade tree were selected. In each farm two fig trees were used for the observations. Fifteen coffee trees under the canopy of the fig trees were marked with permanent pegs. A similar number of trees were marked outside the canopy. Farmers were requested to maintain uniform management practices throughout the coffee farm. The participating farmers were issued with half-kilogram tins and a notebook, and shown how to record the number of tins harvested from the marked areas. Figures for 100-berry weights were obtained by counting and weighing 100 berries of coffee at random during each harvest. Frequent visits were made to the farmers by the extension agent and PLEC team to monitor the progress of the experiment. Soil samples were also taken from the various marked areas and subjected to a full soil chemical analysis. The work was done during the main coffee crop season of October 1999 to January 2000 and the minor crop season of April to June 2000. During that time, the number of participating farmers was increased from three to six. However, because of poor rainfall, observations were only made of three farmers during the main season and two farmers in the minor season.

Yields

The mean yields of coffee under the fig trees and outside the canopy for the two seasons are tabulated in Table 8.1.

There was no significant difference ($p = 0.05$) between the yield of coffee under the fig tree canopy and away from it. However, both the

Table 8.1 Mean yields per plant, 100-berry weight, and half-kilogram tin weight from under and outside the fig tree canopy

	Main crop season (October 1999– January 2000)	Minor crop season (April– June 2000)	100-berry weight (gm)	Half-kilogram tin weight (gm)
Under fig trees	35.4 kg	9.4 kg	171	451
Outside fig tree canopy	28.8 kg	8.0 kg	165	447
Rainfall	821.6 mm	219 mm		
Coefficient of variation (%)	43	35		

farmers and the researchers detected a consistent but small difference. Every 15 coffee trees under the fig trees added 6.6 kg to overall yield during the main season, which (considering the fig trees do not involve additional input) translates into extra income.

During both seasons, yields from coffee trees underneath the fig trees were somewhat higher than from those outside the canopy (22.9 per cent and 17 per cent higher in the main and minor seasons respectively). The biophysical explanation for these consistent but small differences is complex. In the minor season, fig trees shed their leaves and so there is little differential competition for light and water. The protection in this season afforded by the fig tree gives a small increase in yield. Then in the major season, the coffee trees benefit from the mulch and additional fertility provided by the previously shed leaves, again giving a small increase in yields.

Both the half-kilogram tin *kasuku* weight and 100-berry weight were numerically higher for berries harvested under the fig tree canopy than from outside, indicating that the fig tree has a relative positive influence on coffee yield even though statistically it was not significant with the data collected during this study.

During the main crop season the area received only 290.4 mm of rain, which was far below the 821.6 mm received during the previous main crop season. As a result, only three out of the six participating farmers had any coffee harvest. Two out of these three farmers only picked coffee under the fig tree canopy and nothing outside. During the minor crop season only two farmers out of the six participants had some coffee beans to harvest since the coffee had not recovered from the previous drought. On one of the farms there was no harvest outside the fig tree canopy.

Soil analysis

The soil analysis (see Table 8.2) indicated low levels of organic carbon according to the broad ratings of Landon (1984). However, there was no

Table 8.2 Average chemical composition of the soils under and outside the fig tree canopy

	Phosphorus (ppm)	Nitrogen (%)	Carbon (%)
Under fig trees	20.8	0.24	2.73
Outside fig tree canopy	16.7	0.24	2.76
Coefficient of variation	27	6	9

significant difference ($p = 0.05$) between the soil under the fig trees and that from outside the fig tree canopy.

In general, the soils had adequate levels. of available phosphorus according to Landon (1984), ranging from 13.33 to 24.3 ppm. Soil under the fig trees generally had a higher concentration of phosphorus than away from the fig trees. According to Landon (1984), moderate levels of total nitrogen (0.19 to 0.27 per cent) were found, but the levels under the fig trees and away from them were not significantly different ($p = 0.05$).

Farmer feedback

During a group meeting, two major concerns were raised in regard to use of the fig tree.

Some farmers felt that the micro-environment under the fig trees would encourage coffee rust, coffee berry disease, and coffee thrips (*Diarthrothrips coffeae*). Other participating farmers, however, reported that this was avoided by trimming away lower branches of the fig trees and leaving the canopy high over the coffee trees. Similarly, by bending the coffee trees outwards in order to open up the coffee trees, humidity was reduced. They also observed that thrips did not infest under shade. With regard to coffee berry disease, at the time when the weather is cold and conducive to an outbreak (July and August) the fig tree will have shed its leaves. The shading effect will therefore be minimal.

Availability of fig tree seedlings was another concern raised by some farmers. One of the participating farmers was willing to assist in the acquisition of the seedlings.

Farmers listed the benefits of the fig tree generally as provision of firewood, fodder, mulch, and soil fertility replenishment. They also perceived that the fig tree "brings the water up".

The effect of the fig tree on the apparent good performance of coffee could be as a result of several factors. It might be deeply rooted and draw water from a great depth. This water will be in high concentration in its roots and will diffuse into the surrounding soil for uptake by the coffee roots. Another way the fig tree may be benefiting the coffee is by shedding its leaves, which act as a mulch that preserves moisture. Decomposition of the mulch adds organic matter to the soil, resulting in higher coffee yields. The fig tree shading also reduces evapo-transpiration of the coffee plants. This could explain why during the drought of 2000, coffee berries only developed on trees under the fig trees in two of the three farms. The effect of these factors combined is possibly bringing about this apparent yield increase and soil improvement.

Conclusion

The fig tree appears to be beneficial since it provides shade during drought, although during a cold season it could encourage diseases, especially coffee berry disease. However, as the Nduuri farmers observed this does not appear to be a problem since during the cold season the fig tree sheds its leaves.

Given the many other uses of the fig tree and the demonstrated yield increase – especially during drought – the farmers could be advised to plant or retain fig trees. In the process farmers will also be contributing to the conservation of biodiversity. There is a need to make further observations of all six farms involved in this study, since the 2000 drought had a negative impact on the crops.

REFERENCES

ICRAF. 1992. *A Selection of Useful Trees and Shrubs for Kenya. Notes on the Identification, Propagation and Management for Use by Agricultural and Pastoral Communities.* Nairobi: International Centre for Research in Agroforestry.

Jaetzold, R. and M. Schmidt. 1983. *Farm Management Handbook of Kenya*, Vol. 2, Part C. Nairobi: Ministry of Agriculture.

Landon, J. R. (ed.). 1984. *Booker Tropical Soil Manual: A Handbook for Soil Survey and Agricultural Land Evaluation in the Tropics.* New York: Longman.

9

The role of livestock in soil fertility and biodiversity change in Nduuri, Embu, Kenya

John N. N. Kang'ara, Ezekiah H. Ngoroi, Julius M. Muturi, Seth Amboga, Francis K. Ngugi, and Immaculate Mwangi

Introduction

Nduuri is situated on the south-east slopes of Mount Kenya in the agricultural ecological zone (AEZ) Upper Midlands 2, the main coffee-growing zone. Population pressure has led to subdivision of land to such an extent that over 50 per cent of households live on less than 1 ha of land. Only 13.5 per cent of households have 2 ha or more. About 69 per cent of households cultivate their own land, while 29 per cent cultivate undivided family land and about 2 per cent work on rented land. The most common land-use type has been a coffee monocrop with occasional *Grevillea* trees to provide shade. Coffee has been the main cash crop, while a wide range of food crops (including maize, beans, bananas, and tuber crops – cassava, Irish potatoes, and sweet potatoes) is grown, mostly intercropped. Livestock keeping is practised by the majority of farmers in Nduuri, as livestock are associated with social status and provide a source of food and income.

In the recent past farmers in Kenya have been having problems with both dairy and coffee production, which in turn has affected the way of living and land use. There was a need therefore to identify these changes clearly as they affected the biophysical and socio-economic environments, the role of livestock in the dynamic land-use system, and the way forward in protecting on-farm biodiversity and supporting rural livelihoods. A study was undertaken in July 2001 to elucidate, in particular, the role of livestock in a changing economy.

A team of six scientists including a vet, agronomists, and livestock experts were involved in a survey. Fifty-one representative households were randomly selected from the nine villages of Nduuri sub-location. A senior member of each household was interviewed alone or with his or her spouse using a semi-structured questionnaire. A farm visit was made to verify the biodiversity and also to assess the condition of soil, crops, animals, people, and house structures. These were listed, scored, and recorded. The information was entered into a computer and analysed using SPSS (Statistical Package for the Social Sciences) software.

The role of various farm components in the household

Role of coffee

Agriculture is the main source of household income, with 68 per cent of the households interviewed relying solely on farming for their livelihood. Off-farm employment contributes little to Nduuri households. Of those with income from other sources, 12 per cent had a steady income and constituted mainly school teachers and retired pensioned civil servants. The bulk of income came formerly from coffee as the main cash crop. This allowed people to construct good permanent and semi-permanent houses. Education and hospital fees were also met easily by profits from coffee as a farmer would collect a cheque in advance to pay the school and hospital fees and the money was then recovered from sales. Coffee directly provided food security during the drought as needy members used their profits to pay for food delivered by the farmers' cooperative. Most domestic needs were also met by the income generated from coffee.

Similarly, coffee also provided funds for investment in both non-agricultural and agricultural enterprises. Although farmers often had to wait some time to receive payment for coffee (payments were often only made three or four times a year), the amount was large enough to enable them to make investments without a loan. Most dairy cattle were acquired through coffee revenue. Poultry production also has a high initial capital investment, and this too in many households was made possible through coffee sales. The number of coffee trees also indicated the potential to repay loans and they could therefore be used as collateral. These profits from coffee also assisted other farm enterprises, such as providing feed and veterinary services for livestock and fertilizers, seed, pesticides, and labour for food-crop production. Coffee was, therefore, directly and indirectly, central to local people's livelihoods. The majority of production resources (such as land, manure, fertilizers, pesticide, and labour) were consequently devoted to coffee production. In most cases, it received more than two-thirds of the farmyard manure generated.

Role of food crops

Food crops grown by many households provided food security and ease of survival while waiting for payments for coffee. As long as there was adequate rain, production of such crops helped to provide a cheap balanced diet for the family throughout the year. This also reduced dependence on coffee as a source of every household need. This sector also provided a large quantity of herbage in the form of crop by-products or residues for ruminant feeding. However, food-crop production received little of the available production resources. In many farms a large proportion of land was under coffee, leaving little for food crops. They also received less than one-third of manure generated on the farm and often only the residue once coffee-growing land had been treated. The major food crops include maize, bananas, beans, cassava, sweet potatoes, yams, and vegetables. They are usually intercropped among themselves, but farmers have started to intercrop them with coffee trees. Quality scores for the various socio-economic conditions and farm enterprises are presented in Table 9.1. During the survey, maize (which is the major food crop) was scored as a representative of the food crops and coffee for cash crops. Among the farms under study only 22 per cent had a good to excellent maize crop. The others were classified as either fair or poor. In 33 per cent of the farms visited, maize was found to be poor or very poor. Other crops (except bananas and arrowroot, or *nduma*) had similar scores to maize.

Table 9.1 Quality scores (percentages) for various farm socio-economic conditions and enterprises

Description	Maize	Coffee	Household welfare	Livestock	Condition of people
Very good	7.6	3.9	3.9	15.8	3.9
Good	13.4	51.0	17.6	31.4	23.5
Fair	46.1	27.5	60.8	41.2	52.9
Poor	25.3	17.6	17.6	4.8	15.7
Very poor	7.6	nil	nil	6.8	4.0

Table 9.2 Distribution of livestock species

Species	Frequency (as % of all households)
Chicken	71.0
Cattle	69.0
Goats	36.4
Rabbits	20.0
Sheep	15.7
Ducks	2.0

Table 9.3 Distribution of dairy breeds

Breed	Popularity (as % of all cattle)
Cross-breed	21.6
Ayrshire	15.7
Friesian	11.8
Jersey	11.8
Guernsey	7.8

Role of livestock

Livestock were second to coffee in order of importance. Ninety per cent of the farmers in Nduuri had at least one type of livestock. The most popular animals were chickens, which were found in 71 per cent of households. About 70 per cent of households had a flock consisting of three or less hens, one or two cockerels plus around seven chicks. Chick mortality is high because of frequency of disease under free-range management and the incidence of predators. Chickens were followed by dairy cattle, which were owned by about 69 per cent of the households (see Table 9.2). The most popular dairy cattle breed is a cross-breed between exotic breeds or between the Zebus and exotic cattle. The Ayrshire is the most popular of the pure-bred dairy cattle (Table 9.3). The number of cattle on a farm is dictated by holding size: the larger the farm the more feed that is available and the more animals that are kept. Table 9.4 summarizes typical herd sizes in Nduuri. The majority of farms had one or two cattle.

Despite the importance of livestock production, only a little land had been allocated for fodder or pasture production. In about 45 per cent of the farms no land was spared for livestock. In such farms, animals are maintained on food-crop by-products such as maize stover, banana pseudo stem and leaves, bean straw, sweet potato vines, weeds, and multi-purpose fodder trees. Table 9.5 summarizes the percentage of households and the land they have spared for livestock. Most fodder

Table 9.4 Number of cattle per farmer

Number of cattle	% of households
Nil	29.4
1	29.2
2	17.9
3	2.0
4	7.8
5	2.0
6–9	10.7
10 or over	1.0

Table 9.5 Land allocated for livestock fodder production

Land spared for livestock fodder production (m^2)	Frequency (% of the farms)	Cumulative %
Nil	45.1	45.1
1–1,000	9.9	55.0
1,001–2,000	15.7	70.7
2,001–3,000	13.8	84.5
On terraces and boundaries only	15.5	100.0

crops (the main one being Napier grass) are grown on terraces and in small portions not exceeding one-fifth of the farm. Small ruminants are usually tethered on roadsides or in the homestead, or are stall-fed. All cattle are zero-grazed except draught cattle, which are semi-zero-grazed. In most cases, chickens are free-range except during the flowering of low-laying food crops or when they are likely to destroy vegetables like kale and spinach. The birds therefore fend for themselves most of the time – the only regular care they require is provision of grain at some point during the day.

Chickens are mainly kept for meat, eggs, and sales. As they were typically free-range (and hence wander freely), the amount of manure that could be used from poultry production is low. However, the little manure collected from the pen where they were housed at night was used as fertilizer for banana or coffee crops. The eggs provided a regular income, which was used to meet small-scale domestic needs. Mature birds were sold to meet slightly larger domestic needs such as purchasing cooking fat, sugar, pesticides, and even casual labour. Chicken meat also helped in the improvement of family nutrition through a regular supply of easily accessible quality protein.

Sheep were not popular in Nduuri, with only about 16 per cent of households keeping them. Goats were more common and were kept by about 36 per cent of farmers. Reasons for this include the fact that goats

Table 9.6 Cattle manure generated per season

Cartloads per season	Tonnes per season	Frequency (%)
0–9	0.9–1.5	21.6
10–19	2.5–4.0	23.6
20–49	4.5–10.0	11.9
50 or more	12.0–20.0	7.9

do not compete with cattle for pasture, as most were maintained on weeds and indigenous fodder trees such as *Bridelia micrantha*, *Trema orientalis*, *Vernonia lasiopus*, and *Lantana camara* which grow wild on uncultivated niches. Their meat is also preferred to sheep meat and many people eat it as a ceremonial dish. A few dairy goats have been introduced in the area, especially by those with smaller pieces of land and business-minded people who have discovered a large and untapped market for dairy goats, which hence fetch high prices. Both sheep and goats were kept for meat, sales, manure, and ceremonies. They are considered more prolific and multiply faster than cattle, are easy to dispose of, and require less initial capital outlay. They also produce more manure than other smaller stock. The income generated from these small ruminants provides for major domestic needs such as fees, clothes, and purchase of fertilizer and seeds. This is currently more dependable than the income from coffee.

The zero-grazing system of managing dairy cattle has made dairy keeping the major farmyard manure generator in the farms. In some farms milk was the secondary product, the primary product being manure. About 65 per cent of households generate between 0.9 and 20.0 tonnes of farmyard manure each season. Table 9.6 summarizes the quantities of cattle manure produced by different households per season. Crop residues (which could not be eaten by cattle) and any other trash in the farm are collected and mixed up as bedding in the zero-grazing unit. Towards the end of the dry season, manure is removed from the stall and heaped to decompose before being applied to crops. This manure may be combined with inorganic fertilizers before being applied or used on its own. Inorganic fertilizers are rarely used without manure.

Calves provide another income source for the family. Milk payments were (and still are) more regular and reliable than coffee sales. This money is therefore used to pay school fees or supplement the coffee income. Like coffee sales, dairy income is also used to support other productive farm enterprises through purchase of variable inputs such as seed, pesticide, feed, and labour. When income from coffee is not available, households have covered expenditure on school fees or hospital bills by selling cattle. Ownership of dairy cattle and indeed other livestock contributes to household food security both directly (as food) and indirectly (through revenue used to buy food and manure which improves soil fertility and productivity). Livestock also enhance the welfare and the status of the household.

Effect of livestock change on people, land, and biodiversity

Livestock influence biodiversity through their role in the nutrient cycle. Since most nutrients flow from food crops to livestock in the form of crop residues and weeds and back to the crop as manure, removing livestock breaks this cycle. Since the deterioration of the coffee industry in Nduuri, many animals have been sold to pay for urgent domestic needs which were previously met by coffee income. The collapse of dipping services since 1992 and the drought of 2000 resulted in the loss of many dairy cattle through tick-borne diseases and lack of feed. The 30 per cent of the households who did not have cattle (in Table 9.4) are among those who lost their animals during this bad period. The cycle has therefore been broken and adverse changes are now manifest.

The reduction in availability of manure has resulted in a change in biodiversity, simply because species that require fertile soils are losing out to those which can survive in less fertile conditions. Some of the valuable wild species that are disappearing include *Amaranthus* spp., (*terere*), *Solanum nigrum* (*managu*), and *Pennisetum clandestinum* (*kikuyu* grass) which were common in land rich with organic matter or fertilized with manure. They are being replaced by *Rhynchelytrum repens* (or "poverty grass"), *Digitaria scalarum* (couch grass), and *Digitaria ternata*. Both food and cash crops are showing signs of declining productivity from reduced soil fertility. They include vegetables such as kale, spinach, cabbages, tomatoes, and carrots, maize, Irish potatoes, bananas, and coffee.

The effect of reduced manure on plants is lowered biomass production, which is consumable as leafy vegetables and as livestock feed, reduced fruit and tuber size in bananas and potatoes, and low grain and berry yield in maize and coffee. This has not only affected food security, but also the wealth and nutrition status and overall welfare of some families. There is an increase in school dropouts in Nduuri and many people cannot afford good health services. Malaria is common in the area and most of it is resistant to chloroquine. Expensive drugs are therefore required to treat malaria, so many families have planted malaria-curing herbal botanicals and are using them for treatment instead.

Once the livestock-assisted nutrient cycle in a farm ecosystem breaks down, livestock need to be reinstated in their niche in the farm food chain in order to reverse the damage. Coffee payments (which usually involve large sums) are the only way a farmer can pay for dairy cows to replace those sold at a time of need. However, there is no evidence that farmers have yet been in a position to do this and the situation will deteriorate unless they change to other income-generating agricultural enterprises. Already one innovative farmer, Njagi Mbarire, has turned to strategic production of vegetables – harvesting maize while it is green and planting vegetables in its place. The timing of production coincides with the highest market prices. Two self-help irrigation projects, one ongoing and another in the pipeline, have been formed to help in providing water for irrigating such high-value crops.

Conclusion

Livestock play a major role in land-use systems and bring about changes in soil fertility, agrobiodiversity, welfare, and culture in the long term. Efforts to revive the coffee industry (even if the market price improves) need to be supplemented by similar attention in the dairy industry to ensure that the complementary nature of the three farm sectors (livestock, food, and cash crops) is restored. The restocking of dairy cattle should be incorporated in coffee revival packages as policy. If coffee prices continue to deteriorate, farmers should be encouraged to diversify to other lucrative enterprises either by uprooting coffee or intercropping. This calls for change in the coffee-growing policy, which is currently unfavourable to farmers, but is vital if poverty is to be reduced.

This study shows the critical interdependence between different land-use types and farming practices and how a change in one part of the whole land-use system brought on by, say, unfavourable coffee prices can have significant consequences for other parts of the land-use system and for biodiversity. The importance of understanding the whole system when recommending interventions to conserve on-farm biodiversity and to support sustainable rural livelihoods is confirmed. This includes the interdependencies between different crops and their fertility-generating properties, and management practices that rely on the availability of soil-enhancing ingredients.

10

Household diversity of smallholder farms of Nduuri, Embu, Kenya

Charles M. Rimui, Barrack Okoba, Ezekiah H. Ngoroi, and
John N. N. Kang'ara

Introduction

Land-use potential is generally dictated by altitude, rainfall, and soil factors: any change in one of these factors could alter the land-use system of a region. Socio-economic factors also force farmers to try other crops and livestock enterprises to sustain their livelihood. It is through these efforts to overcome the prevailing production constraints that system diversity is introduced. Some crops and livestock are imported into the area despite the production limitations. For instance, areas with coffee-growing potential end up producing crops less suited to the region due to problems with marketing. Culture and traditions also influence what is grown, as do labour and gender issues (which affect resource management). Mean annual rainfall in Nduuri is approximately 1,000 to 1,200 mm with biannual peaks in April and November. Major crops include coffee, maize, beans, and fruit trees. Livestock keeping is also common, although (as with all farming enterprises) production is small scale. The population density ranges from 450 to 700 people per square kilometre.

The production constraints that are exacerbated by the small landholdings have been identified as:

- poor markets for cash crops (coffee and macadamia nuts), hence inadequate income to meet family needs
- declining soil productivity due to continuous cultivation without adequate soil nutrient replenishment

- soil erosion due to cultivation on steep slopes without appropriate soil and water conservation measures
- shortage of livestock feed
- a narrow range of income-generating activities
- labour shortages due to rural-urban migration.

The aim of this chapter is to investigate the socio-economic environment in which the Nduuri community operates, and especially to understand the diversity of farm and family characteristics. The diversity within and between households has been identified in the PLEC agrodiversity framework (see Chapter 2) as a key aspect that encourages other aspects of system diversity. These aspects include agricultural biodiversity (the species and varieties that farmers manage) and organizational and management diversity (the various strategies that farmers use to cope with externally driven pressures such as some of the constraints listed above). Therefore, it is important to understand household diversity as a driver for overall system diversity.

Household interviews and group meetings were carried out with key informants, making use of participatory rural appraisal tools such as using twigs, leaves, and sticks to describe seasonal farm activities' calendars and as a visual aid for discussion. Key informants also carried out a wealth-ranking exercise. They formed lists of household heads and then divided those names according to wealth category. Criteria for developing farmer categories were also suggested.

Group meetings were then used to identify the division of labour of household chores by gender. Secondary data were used to establish how land was acquired and to identify major sources of income by household category, farm, and family characteristics amongst other issues.

Household and farm characteristics

The survey revealed that 92 per cent of the households were headed by males. Household size ranged from two to 25 people, with a mean of 7.6 members. The mean age of household heads was 53.87 years (with a range of 26 to 104 years). The household head's formal education level ranged from none to 12 years. About 85 per cent of the household heads attained at least standard (primary) education, while 15 per cent had no formal education.

Local society is far from homogeneous. Different households have different levels of assets, opportunities, and capabilities. However, categorizing farm households into groups has always been seen as an important component of agricultural research programmes to enable targeting of recommendations to the appropriate group. It may help to investigate questions such as why certain categories of households are wealthier, why some are more successful at growing particular crops or rearing animals, and what constraints each household category faces in terms of agricultural production. Understanding these questions can identify possible solutions, which could help the poorer and/or the less technically able households. But categorization must respect the diversity of households and capture the real differences that affect the management of the farm. Based on the criteria generated by the key informants, four household groups were identified, as shown in Table 10.1.

Land tenure

Land in Nduuri is privately owned. The majority of land owned by Nduuri households was acquired through inheritance, while the rest was acquired through purchase. There is a small portion of community land for schools and other communal facilities such as dispensaries. Land inheritance is predominantly patrilineal, but farms are rarely divided while the father is still active. Sons are encouraged to establish their own farms, and fathers only give land to their sons when they are convinced of their reliability.

Farm sizes and sources of income and expenditure

Across all household groups, holding size varied between 0.1 and 4.0 ha (with an average size of 0.92 ha). In most households (76 per cent) income was wholly derived from the farm. Income had traditionally been generated from coffee and dairy production and from the sale of food crops such as maize and beans. However, poor coffee marketing and inadequate management of local cooperatives had led to decreased returns from coffee sales and farmers paying less attention to coffee production as a result. Main expenditure is on school fees and inputs for crop and livestock production. Food purchase is rare except during seasons where crops have largely failed.

House types and farm tools

Four types of house were observed, the most common being that with a tin roof and mud walls (71.8 per cent). Others were tin-roofed/timber-walled (18.0 per cent), tin-roofed/stone-walled (7.7 per cent), and grass-thatched/mud-walled (2.6 per cent) dwellings. All farmers used hand tools and implements such as *pangas*, flat and forked hoes, knives, axes, and spray pumps for land preparation and crop protection.

Table 10.1 Household characterization in Nduuri

Criteria/farmer group	Group I	Group II	Group III	Group IV
Farm size in ha (average)	1.6–4.0 (2.92)	0.8–3.2 (1.70)	0.4–2.8 (1.48)	0.1–2.4 (0.60)
Number of dairy-grade cows	3–6	1–3	1–3	Few with 1–3
Use of certified seed?	Yes	Yes	Yes	No
Use of fertilizer?	Yes	Yes	Yes	Majority at low level
Ability to educate children	Secondary/university	Secondary/university	Majority secondary	Majority primary
Off-farm income as % of total income	75	50	0	0
% by group (n = 44)	6	18	23	53

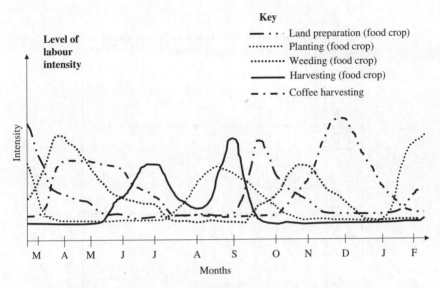

Key

— · · · Land preparation (food crop)

· · · · · · · · · Planting (food crop)

· · · · · · · · · Weeding (food crop)

———— Harvesting (food crop)

— · — · Coffee harvesting

Level of labour intensity

Intensity

M   A   M   J   J   A   S   O   N   D   J   F

Months

Figure 10.1 Labour calendar for major activities

Farm activities' calendar

Predominant farm activities include land preparation, planting of maize and beans, weeding, coffee harvesting, and the harvesting of beans and maize. As Figure 10.1 illustrates, farmers are busy throughout the year, with the minimum labour requirement occuring during the months of June, August, and January. Labour demand for food crops is highest (June to November), while coffee demands the least labour. Demand for labour for coffee production arises in two periods of the year, between October and January and between April and May.

Labour distribution by gender

The family is the main source of labour in all households. However, 76.9 per cent of farmers hire extra labour for specific activities such as land preparation, coffee pruning, planting, and weeding. Coffee takes up 86 per cent of hired labour. In the analysis of gender division of labour, it was found that there were specific activities that could be divided according to whether they were carried out by men, women, or male/female children. These are summarized in Table 10.2. Only activities that concern land management diversity were analysed. The table shows that both men and women are highly involved in land preparation, maize and bean

Table 10.2 Gender division of labour for household activities

Activity	Men	Women	Girls (>15 yrs)	Boys (>15 yrs)	Girls (<15 yrs)	Boys (<15 yrs)
Land preparation (digging)	6	5	3	4	1	2
Planting maize	6	6	3	3	1	1
Planting beans	1	6	5	1	4	3
Weeding in maize	6	1	3	5	1	2
Weeding in beans	1	6	5	1	4	3
Weeding in coffee	6	5	2	1	3	4
Spraying crops (where applicable)	5	2	3	6	1	2
Spraying coffee (where applicable)	6	1	3	5	1	2
Harvesting maize	5	6	4	3	1	2
Harvesting beans	1	6	5	1	4	3
Harvesting coffee	6	5	3	4	1	2
Making SWC structures	6	1	2	5	1	3
Purchasing farm inputs	6	5	3	4	1	2
Marketing coffee	6	5	4	3	1	2
Marketing crops (maize/beans)	5	6	4	3	1	2
Milking and marketing milk	5	6	4	3	1	2
Manuring the farm	6	1	3	4	1	2
Feeding/watering livestock	5	6	3	4	1	2

Note: Scores indicate level of involvement on a scale of 1 (least active) to 6 (most active).

planting, harvesting, and marketing, coffee weeding, spraying, and marketing, purchase of farm inputs, feeding cattle, milking and marketing of milk, and application of manure in the farm.

Nevertheless, though adults performed many of the duties, youngsters also helped in weeding and coffee harvesting, which occurs during the school holidays. Duties exclusive to males are spraying coffee and other crops, muck spreading, construction of soil and water conservation (SWC) structures, and weeding of maize, while exclusive duties for females are planting, weeding, and the harvesting of beans. Older boys are also involved in the duties carried out by their fathers, while older girls help their mothers with household chores.

Conclusion

Despite a reduction in the amount of land under cultivation per household in Nduuri because of population pressures, there is evidence that farmers are not only surviving but are also managing a considerable diversity. Indeed, population pressure may well be a driving force for agrodiversity, causing farmers to diversify as well as intensify production. Most farmers (76 per cent) grow crops for consumption rather than for the market, yet the majority depend on farm crops for income generation. Due to poor market prices, low purchasing power, and institutional management problems, farmers say that they are diversifying to meet their family food and income requirements. Diversification is a strategy that underwrites household security, and household diversity is confirmed as one of the principal ways in which overall diversity is maintained, livelihoods are secured, and external pressures withstood. Contrary to the standard wisdom that population increase leads to loss of biodiversity, Nduuri is a good example of a place where agrodiversity is benefited by externally driven processes of change.

11

Socio-economic factors influencing agricultural biodiversity and the livelihoods of small-scale farmers in Arumeru, Tanzania

Essau E. Mwalukasa, Fidelis B. S. Kaihura, and Edina Kahembe

Introduction

Agricultural biodiversity, or what the PLEC project calls "agrodiversity", is affected by many factors, both from within the farm and household and from external forces. The term "organizational diversity" in the PLEC agrodiversity framework (see Chapter 2) refers to the variety of ways in which farms are owned and operated, and in the use of resource endowments. Exploratory surveys had already indicated that this aspect of diversity appeared to account for a large degree of the variety found between farms. Explanatory elements of "organizational diversity" include labour, household size, resource endowment, and reliance on off-farm employment. Also included are the age groups and gender differences in allocation of farm operations, dependence on the farm as opposed to external sources of support, spatial distribution of the farm, and differences in access to land. This chapter discusses how socio-economic aspects influence the agrodiversity and livelihoods of small-scale farmers in the two distinct PLEC sites of Arumeru district, the semi-arid Kiserian and the sub-humid Olgilai/Ng'iresi demonstration areas.

This study was conducted using the PLEC methodology guidelines (Brookfield, Stocking, and Brookfield 1999) alongside some additional participatory socio-economic techniques in order to analyse organizational diversity. This included cash flow analysis to monitor liquidity.

Feedback meetings were organized to discuss farmers' own recorded findings. Additional data were collected through both informal and formal surveys. Crop market prices monitoring was also done in order to examine local market opportunities for various crops grown in the district.

Farm size

As Table 11.1 summarizes, mean farm size decreased on both sites between 1988 and 1999. Average farm size in the low-altitude, semi-arid zone was significantly higher than in the higher-altitude, sub-humid zone in 1999. According to key informants, decreasing farm size was a reflection of increased population pressure arising from increases in family size and in-migration to both sites.

Land ownership and use intensity

Households in the higher-altitude zone had significantly more farm plots than those in the low-altitude zone, with an average of 3.3 and 1.8 plots respectively (Table 11.2). This may indicate population pressure at Ng'iresi/Olgilai, with households owning many small and scattered plots

Table 11.1 Farm size ownership per household (n = 34)

Zone	Village	Mean farm size (hectares) in a year	
		1988	1999
High altitude	Ng'iresi	1.9	1.1
	Olgilai	1.2	1.1
Low altitude	Kiserian	5.1	2.8
Significance ($p = 0.05$)		0.49	0.02

Table 11.2 Mean number of farm plots per household in 1999 (n = 34)

Zone	Village	Plots/household
High altitude	Ng'iresi	2.6
	Olgilai	3.3
Low altitude	Kiserian	1.8

Table 11.3 Households experiencing food shortage by zone (n = 29)

Zone	Village	Food shortage (%)	No food shortage (%)	Total (%)
High altitude	Ng'iresi	27.3	0.0	27.3
	Olgilai	27.3	3.0	30.3
Low altitude	Kiserian	42.4	0.0	42.4
Total		**97.0**	**3.0**	**100.0**

of land in response to land scarcity and where clan inheritance is the common mode of land acquisition.

Farmers' perceptions of food insecurity at household level

Nearly all farmers (97 per cent) experience food shortage at some point during the year (Table 11.3). Of those, 42.4 per cent live in semi-arid Kiserian (low-altitude zone) and 27.3 per cent each live in Ng'iresi and Olgilai (sub-humid zone). However, farmers indicated that food shortage is most prevalent between planting and harvesting times. Only a few reported food shortages throughout the season, but almost all had to buy food at some time of the year in response to short-term deficits.

Availability and use of forage

Homesteads provided the majority of grazing for livestock in the high-altitude areas (23.3 per cent and 16.7 per cent in Oliglai and Ng'iresi respectively). Meanwhile, those in the low-altitude zone typically used communal land for grazing (see Table 11.4). Distant support plots were used by only a minority of farmers in either zone.

Table 11.4 Main sources of forage for cattle (n = 30)

Major sources	High altitude		Low altitude	Total (%)
	Ng'iresi %	Olgilai %	Kiserian %	
Homestead	16.7	23.3	3.3	43.3
Distant support plots	3.3	6.7	9.4	19.4
Communally owned land	3.3	0.0	34.0	37.3
Totals	23.3	30.0	46.7	100.0

Table 11.5 Summary of farmers involved in cash flow analysis

| | Wealth category | | | |
	Poor	Medium	Rich	Total
High-altitude zone	2	9	6	17
Low-altitude zone	3	7	3	13

Cash flow analysis

Thirty farmers recorded their daily cash income and expenditure (17 in the high-altitude zone and 13 from the low-altitude zone). The sample included farmers from rich, medium, and poor wealth categories (see Table 11.5). However, in the low-altitude zone only the poor were able to provide complete sets of data on cash flow.

The analysis revealed that cash income from off-farm activities was highest for medium-wealth farmers in the semi-arid zone, reaching a peak in January. Cash income from crops was highest between July and October, while income from livestock peaked in February and November. In the high-altitude zone, medium-wealth farmers reported the importance of both off-farm and on-farm activities, while those in the low-altitude zone tended to rely more on the latter.

Crop and livestock production was also an important source of income for rich farmers in the low-altitude area (Figure 11.1). Off-farm and other sources of cash income were the least important. However, rich farmers in the high-altitude zone (Figure 11.2) reported off-farm activities as important income sources, as was crop production, which was at its peak in November and December. The poor in the low-altitude zone relied heavily on off-farm and other sources for cash income, particularly between September and January.

Family need was cited as the main reason for cash expenditure across social classes and agro-ecological zones. Investment in crop and livestock production was typically only possible for rich farmers in the high-altitude zone (Figure 11.3), with peaks in March and October. For the same category in the low-altitude zone (Figure 11.4), cash expenditure for crop production rose only between February and April. Spending on miscellaneous activities (other than for livestock, crops, and family needs) was significant between September and December.

Meanwhile, medium and poor farmers in both altitude zones invested far less money in crop and livestock production than their rich counterparts, and such expenditure was the least significant item households spent money on.

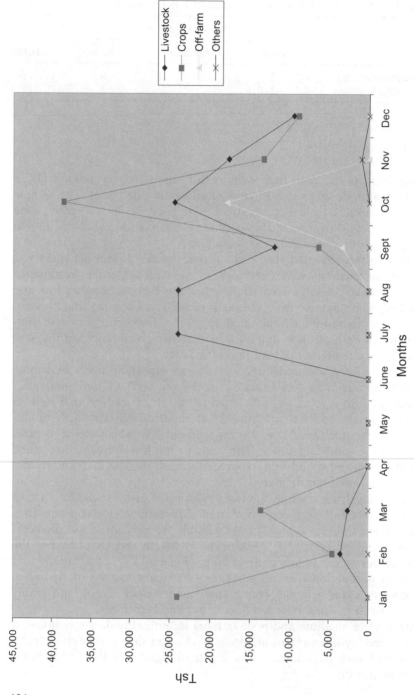

Figure 11.1 Cash income for "rich" households in the low-altitude zone (Kiserian)

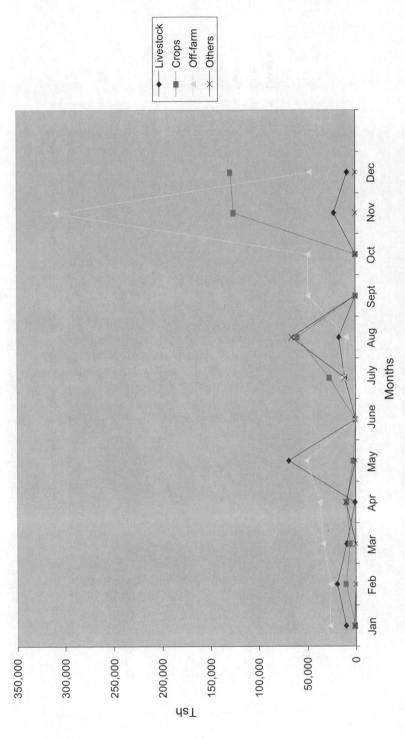

Figure 11.2 Cash income for "rich" households in the high-altitude zone (Olgilai and Ng'iresi)

125

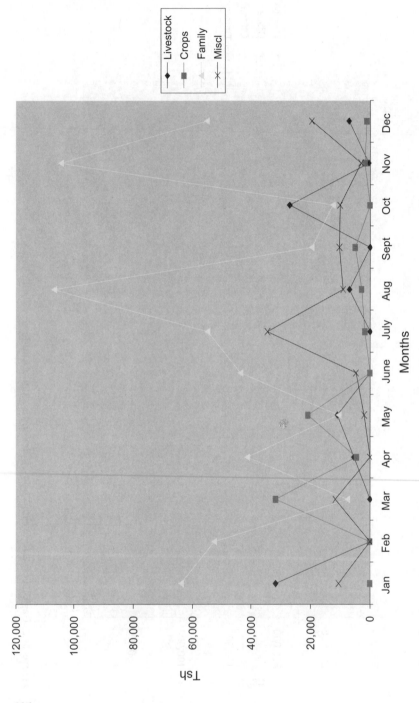

Figure 11.3 Expenditure for "rich" households in the high-altitude zone

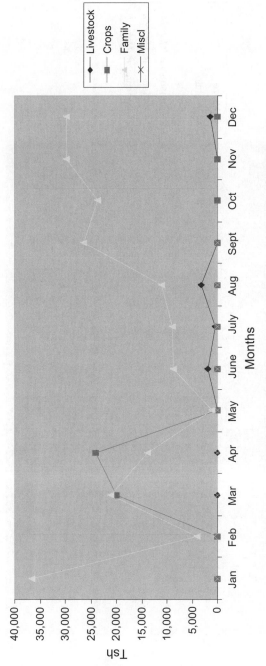

Figure 11.4 Expenditure for "rich" households in the low-altitude zone (Kiserian)

127

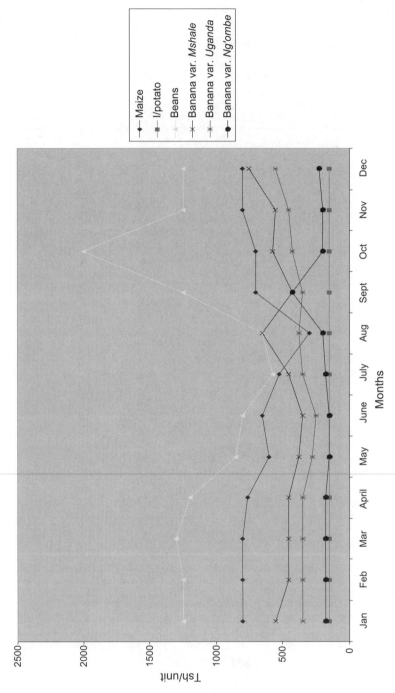

Figure 11.5 Average monthly market prices for various field crops at Chekereni in the 2000–2001 season

Table 11.6 Market risk of commonly grown crops

Crop	Market risk	
	Low	High
Irish potatoes	*	
Banana (*Ng'ombe* variety)	*	
Other types of banana	*	
Beans		*
Maize	*	
Cabbages		*
Tomatoes	*	
Onions	*	
Amaranthus	*	

Local market opportunities and risks

In order to analyse the market opportunity and risks for various crops grown in both agro-ecological zones, prices for those crops were monitored on a monthly basis at various local markets. Chekereni was the most popular local market for various crops grown in both agro-ecological zones (see Figure 11.5).

The degree of market risk for the variety of marketed crops is recorded in Table 11.6. Low market risks were typically due to relatively low price variability compared to other field crops (Figure 11.5). However, risks for Irish potatoes, amaranthus, and tomatoes, when sold in packed units, may be masked by changes to pack size while selling price is maintained.

Other land use and management

Management differences of the two tribes

Traditionally, land tenure does not differ between the two tribes of Arumeru (the Waarusha and Wameru). Land is passed on from fathers to children. The family head, either male or female, controls inherited land. By-laws stipulate that those who misuse the land have their allocation given to another family member. Land shortage in Arumeru means that there is a considerable market for land and purchase is possible. In order to avoid conflict, local laws stipulate that both families involved in the sale of land should be involved in the purchase agreement, which is supervised by the head of family and clan.

Forests

Forest land is normally gazetted (i.e. officially notified and demarcated as belonging to the central or local government). Areas such as Meru Forest are the property of central government and supervised legally by the Forest Department. In some areas, trees such as pines, eucalyptus, and *Grevillea* are planted on a commercial basis. Nearby households are allowed to grow some crops in these zones by agreement with the Forest Department, resulting in a *taungya*-type system of land use where farmers have to agree to respect the trees while they cultivate their crops.

Water sources tenure

Responsibility for the management and protection of water sources rests with the district council, the village government, the Water Scheme Committee, collective water users, or private water users such as the Tanzania National Electric Supply Company. They establish by-laws that govern the use of water sources.

Socio-economic factors and agrodiversity

The results of the survey reveal that socio-economic factors have a variety of implications on agrodiversity and farmers' livelihoods. Factors such as accessibility and use of capital, land (farm size per household), seasonal food insecurity, market forces, and off-farm opportunities are major influences on agrodiversity through creating opportunities to diversify, inflicting constraints on farm investments, and forcing farmers into strategies that minimize household risk. In turn, these changes affect livelihoods. Except for the rich, farmers spend very little of their cash income investing in crop and animal production. These low levels of investment (especially for non-labour inputs) have an impact on soil productivity and subsequently influence agricultural biodiversity. However, low investment in external inputs can mean greater reliance on local biodiversity and protection of existing species and varieties, as well as techniques of management.

Low yields and poverty clearly impact on the food security situation of the area and adversely affect livelihoods. To meet family needs, farmers make choices as to which crops to grow or livestock (number and species) to keep and which enterprises to undertake (on-farm or off-farm). Horticultural crops and off-farm labour were the most common choices. Market forces also influenced which crops are grown.

Access to land under situations of population pressure (from increased family size and in-migration) also influences agrodiversity on farms. Farmers may opt either to emigrate, or to intensify production as well as diversifying to spread risk, especially in drought-prone semi-arid environments.

The increase in off-farm activities, such as part-time waged employment and petty trading (especially for poor and average farmers with less access to land and other resources), reflects the changes farmers have made in response to a decreasing ratio of farm size to household. There is an inverse relationship between non-farm employment and farm size (FAO 1988). Non-farm activities are particularly important for households with little or no land.

Conclusion

Low levels of cash investment for agricultural production have potential negative effects on agrodiversity. Farmers respond to market forces that in turn influence the crops and other enterprises they choose to work with, which in turn affects agricultural biodiversity and related aspects of management. Generally, the greater the levels of cash investment, the greater is the diversification of farm activities and crop types, and the greater is the level of agrodiversity. This is, however, far from being a simple association. Some poor farmers also manage their farms in diverse ways. In addition, the diversification in sources of income has an influence on agricultural biodiversity. Off-farm activities are becoming increasingly important, particularly for households considered poor or middle in wealth. There is also an interesting relationship between rural livelihoods and agricultural biodiversity, where the more secure the livelihoods, the greater is the agrodiversity.

There is still a trend of decreasing farm size in relation to size of household. This is due to population pressure, which creates several concerns with regard to agrodiversity. Strategies to cope with household seasonal food insecurity influence decisions regarding the allocation and management of resources. Those farmers with strategies that involve investment in farm activities to cope with food deficits do so often by diversifying their activities, with consequent benefits to agricultural biodiversity.

The introduction of sustainable technologies is therefore indicated as one means to enable farmers to cope with problems associated with population pressure and droughts. These technologies should also enhance biodiversity, which by itself is a risk-aversion strategy (especially for the poor). Strategies for a wider dissemination of developed technical interventions should be laid down, which should involve important stakeholders in natural resource management. Further studies on the potential role and/or effects of off-farm activities on natural resource use management need to be carried out, alongside the monitoring of household-level population dynamics.

Market infrastructure and information for agricultural produce and inputs are important for the farmers to make rational decisions regarding types of crop to be grown or which enterprises and/or production methods to employ. Policies should have a primary emphasis on making use of untapped potential in, for example, local knowledge of the management of complex associations of species. This includes traditional resource management and its use to enable agricultural biodiversity to be protected for future generations.

REFERENCES

Brookfield, H., M. Stocking, and M. Brookfield. 1999. "Guidelines on agrodiversity assessment in demonstration sites areas", *PLEC News and Views*, No. 13, pp. 17–31.

FAO. 1988. *The Impact of Development Strategies on the Rural Poor: Second Analysis of Country Experiences in the Implementation of WCARRD Programme of Action*. Rome: FAO.

12

Agrodiversity of banana (*Musa* spp.) production in Bushwere, Mbarara district, Uganda

Charles Nkwiine, Joy K. Tumuhairwe, Chris Gumisiriza, and Francis K. Tumuhairwe

Introduction

Uganda is noted for its rich heritage in varieties of bananas. It has been a source of pride amongst agriculturists and conservationists, and a major contributor to Uganda's reputation as a "hot-spot" for biodiversity and the need for its protection for future generations.

Banana growing in Uganda dates back several centuries (Karugaba and Kimaru 1999). It has been sustained by the cultural, social, and economic values that the Ugandan people attach to the crop, alongside the favourable climate (with sufficient rainfall) and good soils in banana-growing areas, especially in the central, southern, and south-western regions of the country. Growth of the banana industry has been achieved through expansion of the land area devoted to production. With an annual production of 8.6 million tonnes, Uganda is a leading banana producer and often referred to as a "banana republic" (Rubaihayo 1991). It is also a major consumer of bananas, taking 90 per cent of local production.

The Ugandan banana industry supports many different sectors, including 75 per cent of farmers (Gold *et al.* 1994), traders, transporters, hotels and restaurants, and breweries. It therefore contributes greatly to government revenue, particularly through taxation. Most edible-fruited bananas, usually seedless, belong to the species *Musa acuminata* Colla (which is itself a product of *M. cavendishii*, *M. chinensis*, *M. nana*, *M. ze-*

brina), or to the hybrid *M.* cross *paradisiaca* L. (*M.* cross *sapientum*; *M. acumianta* cross *M. balbisiana*). In Uganda, the most commonly grown are hybrids, generally *Musa* cross *sapientum* and *Musa* cross *paradisiaca*.

According to Karamura (1994), Uganda has over 100 banana cultivars, implying a rich genetic resource. However, increasing population pressure on land resources, socio-economic transformations such as preferences by markets, and biophysical factors (particularly in hilly and mountainous areas) pose threats to the continuing expansion of banana cultivation. They also threaten the substantial agricultural biodiversity around bananas, including the diversity in management aspects. This chapter outlines the rich biological, biophysical, and management diversity recorded in banana production in the Bushwere demonstration site, Mwizi subcounty, Mbarara district.

Agrobiodiversity potential

This study was carried out to establish the agrobiodiversity potential of banana growing in a rugged highland area. Specifically, the objectives were as follows:
- to assess the effects of biophysical diversity on banana biodiversity
- to inventory the biodiversity found in banana-based field types
- to establish the management diversity of banana cultivation
- to assess the suitability of banana-based cropping systems for promoting agrobiodiversity conservation.

In order to do this, a sample area was selected considering landscape diversity, age of plantation, and cooperation of participating households. Agrobiodiversity assessments were then carried out on 20 m² plots (or the entire plantation if the plot were small). The assessment included identification of clones by local expert farmers, and counting stools of each clone and other components as outlined in the biodiversity advisory group (BAG) guidelines on 5 m² and 1 m² plots.

Informal discussions were held with household members of participating farmers or owners of the fields to capture the organizational diversity, management regimes, and utility of the biodiversity found in the sample area.

The importance of bananas

Banana is a major food and cash income generator for households in the Bushwere PLEC site area. Every household in the area has at least one piece of land under banana cultivation. This means that as numbers of

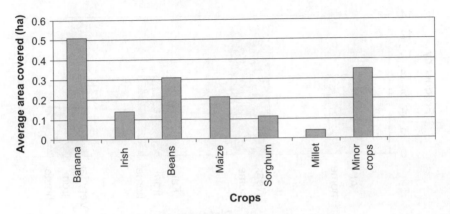

Figure 12.1 Average crop hectarage per household in Bushwere (n = 92)

households increase, more land will be taken up by banana production. The average area devoted to banana per household was 0.5 ha, which is 33.3 per cent of the cultivable land available to each household (Figure 12.1). This is above the national average proportion of cultivable land under bananas, which is 18 per cent according to MPED (1997). This can be attributed to the crop's ability to sustain food supply while also earning household income all the year around. When it comes to earning household income, banana ranks highest (23 per cent), followed by beans (22 per cent), and Irish potatoes (15 per cent). "Minor crops" (19 per cent) include groundnuts, field peas, soya beans, pineapples, tomatoes, cabbages, and sugar.

Spatial distribution

Figure 12.2 shows the distance of plantations from home. The majority (61.8 per cent) of plantations are near home, i.e. within less than 500 metres, while nearly a quarter (23.5 per cent) are "far" from home and a few (14.7 per cent) are "very far", i.e. one kilometre or more from home.

Table 12.1 summarizes spatial distribution of banana plantations by landscape types. It is important to note that banana crops are grown on all landscape types: hilltops, backslopes, and valleys. However, the majority (40 per cent) were found on hilltops, with 37 per cent on backslopes and 23 per cent in valleys.

Valley plantations are typically over 20 years old. However, recent expansions extend to the backslopes and hilltops in response to population growth. Younger families migrate to hilltops and backslopes, and every

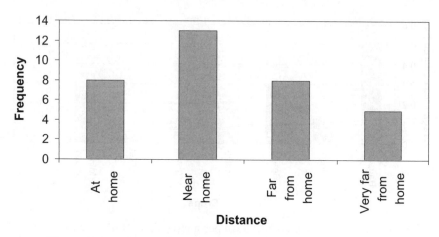

Figure 12.2 Relative distance of the plantation from home (n = 34)

Table 12.1 Banana garden distribution by landscape position (n = 57)

Age groups (years)	Number of plantations			
	Hilltop	Backslope	Valley	Total
≤2	1	4	0	5
3–5	5	2	0	7
6–10	4	1	0	5
11–20	3	5	0	8
21–40	3	6	4	13
<41–45	7	3	9	19
Total	23 (40%)	21 (37%)	13 (23%)	57 (100%)

new household starts a banana plantation, preferably as near the homestead as possible.

Diversity of varieties/clones

Functional grouping

Farmers grouped banana varieties (the terms "varieties", "clones", and "cultivars" are used interchangeably) grown in their area into categories: cooking, beer-making, dessert use, and roasting. These were later matched with genome categorization as summarized in Table 12.2. Cooking varieties were the most popular, followed by juicing/beer-making types. This is reflected on a national level too – in Uganda as a whole,

Table 12.2 Farmers' grouping of banana varieties and their scientific genome categories (n = 57)

Functional group	Utility	Genome	Group name in Bushwere	No. of cultivars
Cooking banana	Steamed (*matooke*) Boiled (*katogo*)	AAA-EA	*Enyamwonyo*	39
Beer type	Juice, beer, and dry gin from ripened fruit	AB ABB AAA-EA	*Kisubi* *Kayinja* and *Musa* *Embiire* (*mbidde*)	1 2 7
Dessert banana	Eaten ripe	AB AAA (*Gros michel*) ABB	*Kabaragara* (*ndizi*) *Bogoya* *Kisamunyu* (*kivuvu*)	1 1 1
Roasting type	Roasted green or ripe	AAB	*Gonje* (*gonja*)	1

Note: Names in brackets are equivalents in Luganda dialect (Karamura 1994).

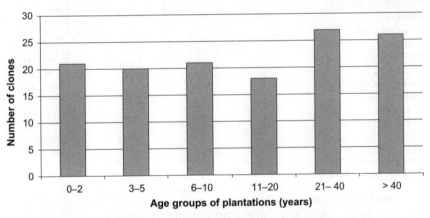

Figure 12.3 Distribution of varieties by age of plantation (n = 57)

cooking varieties account for 96 of the total 131 banana varieties (Karamura 1994).

Distribution of clones in different farms and age distribution of plantations

The survey indicated that all plantations generally have more than 18 varieties, but older plantations (over 20 years of age) have higher number of clones than the younger ones (Figure 12.3). This implies that

Table 12.3 Banana varieties in Bushwere (n = 57)

Banana cultivar (local names)	Average number of stools/400 m^2	Distribution in village
Enshenyi	28	Widespread
Embiire (enkara)	9	Widespread
Kabaragara	4	Widespread
Enzirabahima	4	Widespread
Enjagata	2	Widespread
Embiire (entukura)	2	Widespread
Entaragaza	2	Widespread
Enzirabushera	2	Widespread
Rwamigongo	2	Widespread
Enyaruyongo	2	Widespread
Nyakyetengwa	1	Common
Makunku	1	Common
Embururu	1	Common
Enyarukira	1	Common
Embiire (enyabutembe)	1	Common
Butobe	1	Occasional
Kayinja	1	Occasional
Embiire (engumba)	<1	Rare
Mujuba	<1	Rare
Kaitabunyonyi	<1	Rare
Bogoya	<1	Rare
Nyakinika	<1	Very rare
Burikwezi	<1	Very rare
Embiire (engoote)	<1	Very rare
Kisamunyu	<1	Very rare
Musenene	<1	Very rare
Enzinga	<1	Very rare
Gonje	<1	Very rare
Katwaro	<1	Very rare

Widespread = found in almost every plantation
Common = found in about 50 per cent of plantations
Occasional = found in 1–2 per cent of plantations
Rare/very rare = found in less than 1 per cent of plantations

younger generations are becoming more selective of which cultivars they grow. Alternatively, it could be an indication that some varieties are becoming rare.

Abundance of different varieties

Table 12.3 shows the varieties encountered in the sample areas of 57 farms and their abundance. Over 50 varieties are known from the Bushwere area. In the sample of 30 varieties found in this survey, 10 are classified as "widespread", i.e. found in almost every plantation, and 12 are "rare" or "very rare". The remaining varieties (shown in Table 12.3) occur commonly or occasionally.

The *enshenyi* cultivar was the most dominant in the area because of its special ability to give high numbers of suckers. Farmers reported that *enshenyi* gave a bunch size which is acceptable both at home and on the commercial market. When choosing which varieties to grow, availability of planting material, acceptability of the bunch size, and quality were the main factors considered. However, a few people chose some cultivars (such as *oruhuna* and *enzinga*) for other cultural reasons.

Management diversity

Management of banana plantations in Bushwere consists of soil and water conservation, mulching, weed management, soil fertility maintenance, and intercropping. These will now be discussed in turn.

Soil and water conservation

Despite the plantations' high susceptibility to soil erosion (due to the steep-sloping nature of the land), many farmers (58.8 per cent) are not practising any form of soil and water conservation. However, there is evidence that some people (33.8 per cent) had constructed soil and water conservation trenches along contours in their plantations. A few farmers used diversion channels and soak pits.

Mulching

Use of crop residues and grass was not common (20 per cent and 18 per cent respectively). Instead, almost all banana plantations were self-mulching (banana leaves and fibres). No mulching was recorded in young plantations, which were still being intercropped with annual crops like beans, maize, and millet. Discussion with farmers revealed that grass and other mulch materials were not available to most people. Even use of crop residues as banana mulch was constrained by the long distances between annual-crop fields and banana plantations. The 20 per cent of farmers who mulched with crop residues were most likely to be those near homesteads, since threshing crops like beans and peas was done at home. The residues were then thrown into the nearby banana plantations.

Weed management

Table 12.4 shows several ways of weed management in banana plantations. Heaping and scattering weeds in the plantation were equally commonly practised (38.2 per cent for each).

Table 12.4 Weed management practices in banana plantations (n = 68)

Weed management practices	Frequency	%
Heaping	26	38.2
Scattering	26	38.2
Burying	4	5.9
Removing from plantation	12	17.7
Total	**68**	**100.0**

Table 12.5 Soil fertility management practices in plantations (n = 68)

SFMP	Frequency	%
Application of crop residues	10	14.7
Use of farmyard manure	3	4.4
Application of domestic rubbish	10	14.7
Use of compost manure	3	4.4
No SFMP	26	38.2
Total	**68**	**100.0**

Farmers reported that heaping was most practical during the rainy seasons and/or to speed up rotting and reduce frequency of stubborn weeds such as *commelina* and couch grass. Scattering was done during the dry seasons when weeds dried quickly and the problem of sprouting was minimal. A few people (5.9 per cent) dug ditches and buried stubborn weeds to turn them into manure.

The microclimate under a banana crop is humid and conducive to vigorous growth of many weeds. Farmers' awareness of this, coupled with their concern to maximize banana yields, meant that weed control was considered a major activity. They also took advantage of this microclimate to conserve other useful species. Less than a fifth of farmers (17.7 per cent) removed weeds from plantations, because most appreciated the fact that dead weeds contribute to manure and mulch, thus recycling plant nutrients to feed the banana crop.

Soil fertility management practices

The majority of farmers (38 per cent) had no means of maintaining soil fertility on plantations (Table 12.5). Application of crop residues and domestic rubbish to the plantations was practised by 14.7 per cent of households. Use of both farmyard manure from livestock sheds and compost was rarely carried out (just 4 per cent of farmers in the study area). Most farmers do not own cattle, and the few who did kept them on

Table 12.6 Major types of intercrops in banana farming and their distribution $(n = 68)$

Intercrop type	Frequency	%
Banana/beans	18	26.5
Banana/coffee	8	11.8
Banana/beans/maize	7	10.3
Banana/beans/Irish potatoes	4	5.9
Banana/coffee/beans/maize	4	5.9
Banana/beans/peas/coffee	3	4.4
Banana/beans/maize/cassava	1	1.5
Banana/fruits, e.g. avocados	1	1.5
Banana/coffee/beans/Irish potatoes	1	1.5
Banana/beans/Irish potatoes	1	1.5
Banana/sweet potatoes	1	1.5
Banana/beans/sugarcane	1	1.5
Total	**50**	**73.5**

an open-grazing system like goats and chickens. They therefore did not have much manure to use. Only a few progressive farmers, including the PLEC demonstrators (James Kaakare, Fred Tuhimbisibwe, and Charles Byaruhanga among others), had learnt to make compost manure from agricultural extension workers, PLEC demonstration activities, or farmers' associations to which they belong. However, even these farmers complained that availability of biodiversity in adequate amounts to make good compost manure was a big constraint.

Intercropping

Although Bushwere appears to be increasingly covered by banana plantations, this study revealed that there were about 12 major intercrop types in the banana plantations (Table 12.6). These make up 73.5 per cent of all banana plantations in the area. Only 26.5 per cent are pure banana stands. Table 12.6 shows that banana/beans was the most common (26.5 per cent) banana intercrop field type, followed by banana/coffee (11.8 per cent) and banana/beans/maize (10.3 per cent).

Reasons for the popularity of banana

Ugandan banana crops are typical of species that have naturalized in introduced niches over a long period of time (approximately 2,000 years). The fact that every household grows bananas reflects the high regard the

Ugandan people have for the crop. The perennial nature and high cultivar diversity increases the crop's resilience to droughts, hailstorms, wind,
and declining soil fertility. The crop therefore plays a major role in food
security and income generation at household level, and subsequently at
national level and in the economy.

The Bushwere demonstration site that was inhabited by people in the
early 1940s has about 54 banana varieties, which are almost equally grown
on every landscape position (hilltop, backslope, and valley). Management
of the plantations is routine – pruning (leaves and fibres), de-suckering,
and weeding. Little is done directly on improving and maintaining soil
fertility or controlling pests (weevils and nematodes) and diseases. Coupled with little in the way of soil and water conservation, this brings
about concerns over the sustainability of production – particularly as
crop production is carried out on steep inclines. Lack of integrated management of plantations can lead to loss of plants, especially susceptible
cultivars of this important crop, and decline in yields as has been reported in central Uganda by Rubaihayo (1991).

Biodiversity in banana plantations

Banana plantations, even at typically low levels of management, support a lot of biodiversity. Deliberate conservation of useful plant species
is mostly done in banana-based field types. The canopy structure in traditional banana plantations reflects the natural multistorey system of
tropical forests, similar to the Chagga home gardens of Tanzania. Most
plantations, especially the well-managed ones, comprise four levels, as
summarized in Table 12.7.

Innovativeness in such a system promotes sustainable agrobiodiversity
conservation and utilization. Some of the Bushwere farmers (including
Kaakare and Muhwezi) were skilful in spatial arrangement of different
crops and plant species conserved in banana plantations, which reinforces
the production system. For instance, several farmers in Bushwere planted

Table 12.7 Levels of canopy structure in banana plantations

Level	Features
One	Tall trees such as castor and avocado, e.g. *mpiirwe*
Two	Bananas and pawpaw
Three	Shrubs like red pepper and eggplants
Four	Short, creeping plants like tomatoes, cocoyams, and amaranthus

Musa cross *paradisiaca* (*kabaragara*) around the edges of the plantation to protect cooking bananas against wind. Wind damage was a common problem in banana production of highland areas. Pseudo stems of *paradisiaca* were stronger than those of *sapienta*, especially the AAA-EA genotype.

Farmers also used other plant species, such as *Setaria* spp. and cocoyams (*amateyere*) to stabilize banks of soil and water conservation structures (trenches and soak pits) in bananas.

This is one of the reasons that the option of integrating stall-feeding of livestock into crop (especially banana) production has been more readily accepted by the PLEC collaborating farmers. This system encourages farmers to grow fodder species within and around the plantation. The by-products of bananas (like male buds and banana peelings) can be fed to livestock. In the case of severe fodder shortages in dry seasons, the pseudo stems can be chopped for cattle feeding. In return, the cattle production system provides farmyard manure, offering better nutrition and consequently higher banana yields. The household benefits from both the livestock and the banana crop in terms of improved nutrition and higher income. An example of such an integrated system in practice is James Kaakare, a PLEC demonstration farmer in Bushwere, who declares that his household's welfare has improved greatly since he adopted the integrated system. He claims that his youngest daughter is bigger and healthier than her elder siblings and, unlike his first two daughters (now married), the younger five children all go to school.

Conclusion

The banana crop offers rich genetic diversity. Banana-based land-use stages have great potential that farmers can capitalize on for agrodiversity conservation and sustainable use. However, the crop requires technical skills in correct management, including spatial arrangements and intercropping alongside integrated pest and soil fertility management.

With the expansion of the banana culture and the national plan for modernization of agriculture, which includes commercialization, efforts to maintain this agrobiodiversity of banana production are more crucial and urgent than ever before. Commercialization of crop production systems can lead to net nutrient export, consequently resulting in drastic collapse of the ecosystem if farmers are not adequately sensitized and facilitated to replenish nutrients with soil inputs. Integration of livestock stall-feeding systems into banana production is a promising option for enhancing the sustainable use of the already rich agrobiodiversity.

REFERENCES

Gold, C. S., E. B. Karamura, A. Kigundu, F. Bagamba, and A. M. K. Abera (eds). 1999. "Monograph on geographic shift in highland cooking banana production in Uganda", *African Crop Science Journal*, Vol. 7, No. 3.

Karamura, D. 1994. "Numerical taxonomic studies of the East African Highland banana (*Musa* AAA-EA) in Uganda", PhD thesis. Montpellier: INIBAP.

Karugaba, A. and G. Kimaru. 1999. *Banana Production in Uganda*. Technical Handbook No. 18. Nairobi: RELMA, p. 72.

Ministry of Planning and Economic Development (MPED). 1997. *The Republic of Uganda 1997 Statistical Abstract*. Entebbe: MPED.

Rubaihayo, P. R. (ed.). 1991. *Banana-based Cropping System Research. A Report on Rapid Rural Appraisal Survey on Banana Production*, Research Bulletin No. 2. Kampala: Makerere University.

13

Change in land use and its impact on agricultural biodiversity in Arumeru, Tanzania

Jerry A. Ngailo, Fidelis B. S. Kaihura, Freddy P. Baijukya, and Barnabas J. Kiwambo

Introduction

Until now, land-use changes, their role in the evolution of the present farming systems, and causes of biodiversity change have been largely neglected by researchers in Tanzania. Land-use changes are influenced by many complex factors. The most dominant forces governing changes in land use are population growth and concomitant demand for land-use products, which differ considerably across and within farming systems.

There have been continued changes in land use in many parts of Tanzania, brought about by diverse climatic conditions, changes in population, land pressure, and changes in socio-economic factors. Many of these changes have been discerned by remote-sensing techniques. Figure 13.1 provides a hypothetical summary of the major causes and effects of animal and human populations on the agro-ecosystems in Arumeru.

Land cover influences biodiversity, soil erosion, or nutrient balances and was therefore considered the most important factor when assessing sustainability on the slopes of Mount Meru. Aspects of biodiversity are critical to the overall aim of promoting a more ecological approach in agricultural systems and the integrated management of land resources with a view to enhancing agro-ecosystem sustainability (Brookfield and Padoch 1994). Farmers are rational in their sustainable use of resources. For example, in the Arumeru district, farmers rotated production across their land in order to utilize land resources successfully and maximize

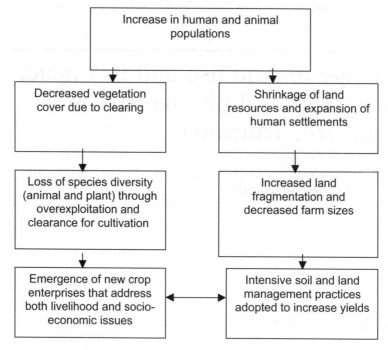

Figure 13.1 Hypothetical summary of land-use and biodiversity changes in Arumeru district – causes and effects

crop yields. Similar cases are reported elsewhere by Padoch and de Jong (1992) in which farmers have concentrated on conservation. Biodiversity conservation has a fundamental role in food provision and the livelihoods of farming communities and the future of people all over the world (Brookfield and Padoch 1994). To small-scale farmers, changes in agro-biodiversity have been an important component in ascertaining their livelihoods. This is because of its importance as the main life-support system.

Most of the data in this study were collected at farm level. The fundamental task was to investigate how small-scale farmers on the slopes of Mount Meru have survived despite the pressure of shrinking land resources. The study aimed to look into the effects of land-use changes (a result of population pressure) and their influence on both agriculture and biodiversity. The area under investigation is the windward slopes of Mount Meru, Arusha, Tanzania.

The study taps into the rich experience of the indigenous people of different age groups and farmer categories, and will assess the sustainability of the system as described by those key informants.

Background

The approximate geographical locations of the study villages are 36°42′50″E; 3°19′36″S and 36°45′00″E; 3°19′36″S. The study villages – Ng'iresi, Olgilai, Moshono, and Kiserian – are along a line of transition from the sub-humid upper slopes of Mount Meru to the semi-arid lower slopes. Most of the PLEC project research activities have been conducted in these villages. The mountain is volcanic and reaches a peak of 4,562 metres above sea level. Volcanic eruptions in the past have resulted in the formation of several cones of various sizes around the mountain. The slopes of the cones are steep and dissected, forming both broad U-shaped and V-shaped valleys. The four study villages are located on the slopes of a volcanic cone called Kivesi. The area is characterized by steep slopes – in some places 30 to 50 per cent – as well as broad valleys and a gently undulating plain extending to Kiserian village. The lithology of the area is late Cretaceous to recent volcanic materials composed of basalts, trachytes, and pyroclastics (Morss 1980). The soils are generally rich in nutrients, but fertilizer application is considered indispensable for better crops.

The projected population size of the Arumeru district for the year 2001 was about 320,000, with a growth rate of about 3 per cent. This rate is one of the highest in the country, slightly above the national growth rate of about 2.8 per cent. This is resulting in an increasing population density, which is currently about 110 persons/km^2 (Bureau of Statistics 1988).

Climate

Long-term rainfall data show that there are two major rainfall seasons. The short rains season normally occurs between November and December, while the long rains begin as early as March and finish at the end of May or June. The second growing period, however, is not reliable. Decreasing total rainfall and high variability between seasons are considered partly responsible for the changes in agrodiversity and are blamed for crop failures. One commonly given local reason for the increasing variability in rainfall is the impact of forest clearances by many farmers along the slopes of the mountain for cultivation and the effect on rain-cloud formation.

Farmer interviews

In each village, 30 farmers of different wealth categories (rich, average, and poor) were interviewed. Questions covered farm productivity and land-use changes, including the adjustment strategies and the range of options available to indigenous farmers and immigrants to adjust to changes. Farmers' responses were important in gathering data on field types and land-use changes in time and space.

Transect walks and historical records

Transects which could be depicted in detail by participants of different age groups (see Table 13.1) and which were considered representative of the area were selected for each village. Key informants accompanied researchers during visits to the selected routes and information given by members of different age groups was recorded. In most cases the information given by older farmers was considered more reliable than that from younger groups.

In each age group, 10 to 15 knowledgeable farmers were selected from each village. The main assumption was that individuals in the same age group have common knowledge and understanding of the various social and other issues which have arisen during their lifetime. It was with this consideration in mind that the farmers from the age groups were interviewed to provide information on land-use changes over time. Information gathered from these farmers was used to supplement information collected from the selected households. The different age groups are presented in Table 13.1.

Information gathered from the farmers was checked against available records and a good correlation was found between the sources. In many cases the information gathered from farmers was useful, sufficient, and relevant for assessing land-use and biodiversity changes.

Table 13.1 Age groups according to the Waarusha tribe in Arumeru district

Age group	Approximate year of birth	Approximate present age
Ilnyangusi	1930	70
Ilseuri	1950	50
Ilmakaa	1970	30
Illandisi	1990	10
Ilameijok	2000	1

Causes of land-use and biodiversity changes

Population

There are many factors that can be postulated to be major causes of land-use and biodiversity changes, but the continued increase in human population (from 238,000 in 1978 to an estimated 430,500 in 2000) is one of the most prominent factors in Arumeru. Population pressure has led to a decrease in the area available for building, grazing animals, and more importantly for cultivation of food and cash crops. Similarly, growth in population has led to increased exploitation of the rich biodiversity for human and animal use. People required trees for building their homes and pasture for feeding their animals. It is such negative interactions of population and environment that have upset the ecological balance in Arumeru.

The local area surrounding the study site villages has about 26 per cent of the total population in Arumeru (Bureau of Statistics 1988). The population density is second only to Arusha municipality and urban area. The district has been facing continuous population growth, which is largely attributable to natural increase. Migration to the district is not considered significant – participants reported that more people are considering leaving the district in search of more land than moving into the district.

Calculations for population growth have taken into account the growth rates as provided by data from the population statistics of Tanzania. The projections made for the year 2000 and beyond also predict further population increase. At this rate of population increase, resources such as soil, water, and vegetation will continue to deplete.

Periodic changes in weather

Most farmers (60 per cent) believe that decreasing rainfall (and in some cases acute drought) has upset the ecosystem. Rainfall data show a high degree of variability between years and this, coupled with poor distribution throughout the year, is considered to be the main problem. This unpredictable nature of rainfall and growing seasons has forced some farmers to abandon older crops, thereby negatively affecting on-farm biodiversity.

Depleting land resources

Farmers have opted to migrate to new areas in response to continued land fragmentation. Sixty per cent of farmers have purchased land elsewhere, and most have almost abandoned the tradition of distributing their land to children as inheritance. In many families, there is very little land left for further distribution to the coming generations.

Effects of population pressure and weather changes in the area

The changes in land use and biodiversity have had a direct influence on the lives of local people. For example, the deterioration of agro-ecosystems has had a direct effect on the livelihoods of the poor because of diminishing production of food crops. The effects of population increase (as suggested by participant farmers) will now be discussed.

Increased land fragmentation

About 90 per cent of farmers reported that land fragmentation was a result of continuous increases in population and had resulted in a decrease in the effective area available for agriculture. Most families have insufficient land for their children. Some have moved to new areas in search of new land to cultivate, while others have opted to intensify production on their existing fields and supplement that production with support plots away from the homestead.

Table 13.2 summarizes farm size according to the wealth category of the participant farmers in the study villages. Large farm sizes were found in semi-arid Kiserian. "Poor" farmers in both sites experienced the biggest constraints regarding land. The data on farm size confirmed the land pressure – especially in the sub-humid uplands and for those considered "poor".

Introduction of quick income-generating crops

Sixty per cent of the farmers included in the study reported that crops such as Irish potatoes, onions, and flowers have been cultivated in re-

Table 13.2 Average farm sizes (hectares) per household in survey villages

Village name	Household category		
	Rich	Medium	Poor
Olgilai/Ng'iresi	0.92	0.72	0.36
Moshono	1.42	1.34	0.60
Kiserian	3.06	2.12	1.53

sponse to market demands. There is a local demand for potatoes, onions, and cabbage within the Arusha municipality. Growing such crops can quickly provide farmers with cash income. This has led to a concentration of effort on the cultivation of such crops, thus limiting the amount of time farmers can spend tending their coffee crops. Flowers are also grown for foreign markets under the supervision of exporting companies which have selected a few farmers to produce flowers for them. Many other farmers are interested in growing flowers in the future.

Over time there has been increasing and decreasing demand for some animal and plant products. Sometimes these changes have led to over-exploitation or abandonment of some plants and animals. For example, the reduction in *ngwala* (*lablab* sp) cultivation by farmers reflects the dramatic fall in demand for *lablab* in the area, with local people choosing other crops as a dependable source of food. However, market surveys carried out during the PLEC project indicate that there is high demand (and subsequently a good price) for *ngwala* in Kenya. The crop has now been reintroduced.

Change in land-use and cropping systems

Table 13.3 presents information on the increase/decrease of certain crop enterprises over the years. Changes have occurred to both the composition of traditional field types and agrodiversity as a whole.

At both sites, maize production is dominant, followed by beans. Bean production slightly declined over time in Olgilai/Ng'iresi, probably due to an increase in potato production as a cash crop. Coffee production also declined, although most of the coffee grown now is intercropped with other crops such as cabbages and potatoes.

It is clear from Table 13.3 that farmers have set priorities for some crops. For instance, maize and beans in both zones have occupied a significant proportion of all crop production, reflecting the farmers' priority of food security. In the sub-humid zone, coffee as a cash crop has occupied a significant (13 per cent) part of the cultivated land. This is because most farms still have coffee intercropped with other annuals, despite its decreasing importance as a cash crop. Irish potatoes and finger millet occupy an average 8.6 and 8.4 per cent of cultivated land respectively. These are not the main food crops, but are used as cash crops.

In the semi-arid zone the predominant cash crop is pigeon peas. Flowers are grown to generate income, but production is typically controlled by the export company/buyer. Maize (24 per cent) and beans (18.2 per cent) are the most popular food crops. It can therefore be concluded that when there is land shortage (as explained above), allocation of land to a crop or combination of crops depends on the ability to improve household food and income requirements.

Table 13.3 Farmers' perception of trends in land use for various crops (percentages) in the study areas (1930–2000)

Zone	Crop	Age groups					
		Ihnyangusi (1930)	Ilseuri (1950)	Ilmakaa (1970)	Illandisi (1990)	Ilameijok (2000)	Average
Sub-humid Olgilai/Ng'iresi	Maize	30	30	25	20	27	26.4
	Beans	25	25	15	15	11	18.2
	Finger millet	20	15	5	2	–	8.4
	Lablab	10	8	5	1	–	4.8
	Cowpeas	5	5	2	1	–	2.6
	Sweet potato	2	2	2	2	2	2.0
	Coffee	1	5	25	22	15	13.6
	Banana	5	6	10	18	20	8.2
	Taro	1	2	2	4	5	2.8
	Irish potato	1	2	8	12	20	8.6
	Cabbage	–	–	1	3	5	2.2
Total		**100**	**100**	**100**	**100**	**100**	
Semi-arid Kiserian	Maize	31	29	20	22	20	24.0
	Beans	15	17	20	23	18	18.6
	Finger millet	11	11	10	6	8	9.2
	Lablab	14	8	5	3	3	6.6
	Cowpeas	9	6	8	5	6	6.8
	Sweet potato	5	1	5	2	9	4.4
	Cassava	4	4	6	8	2	4.8
	Pigeon pea	11	11	10	11	13	11.4
	Chickpeas	–	8	9	10	5	6.4
	Sorghum	–	5	2	3	4	2.8
	Flowers	–	–	5	7	12	4.8
Total		**100**	**100**	**100**	**100**	**100**	

Coping strategies in response to changes in land use and biodiversity

Farmers have devised several strategies to cope with the effects of land-use changes. For those who cannot settle elsewhere in the region, they have maximized the use of their plots with considerations of the varying seasons, market demand, and the crops' ability to address farmers' needs for cash and food. These coping strategies will now be described in detail.

Intensification of land use in the different seasons

Farmers intensify production on available land in accordance to season and market demand. Such a system has allowed the farmers to exploit the scarce resource to meet their economic needs.

The production system in the area is a diverse one, involving the use of a range of available soil and land types. It is also a complex system, since it has the ability to integrate the use of micro-ecological niches within the farmers' area according to the prevalent intra- and inter-seasonal conditions.

Every year both permanent and short-season crops are cultivated. The short-season crops such as maize, beans, Irish potatoes, and most vegetables are observed to be rotated between plots. Tree crops such *Grevillea*, *Pinus*, *Calliandra*, and other fodder trees are grown in the same plots each year and season.

The reasons behind farmers' land-use decisions are as follows:
- declining number of fields has forced farmers to undertake crop rotation
- control of pests and disease
- response to market demands
- land scarcity has forced most of the farmers use their land more effectively, e.g. growing several crop types on one piece of land.

Adoption of improved soil and water conservation techniques

Alongside changes in land use, a large proportion (70 per cent) of the interviewed farmers have introduced soil and land management technologies as a result of external interventions as, for example, with the PLEC activities. These included the use of fertilizer, use of improved seeds, afforestation, and soil conservation, with the aim of assisting farmers to conserve the resource base and improve yield.

Improved breeds of cattle and goats have been introduced and are zero-grazed in the sub-humid zone. This increased the need for simple but effective technologies in the light of the population pressure explained above. To this end, grassed contour bunds were introduced and maintained in order to curb soil erosion. The grasses were planted on contours and were cut and carried home to feed the animals. Similarly, soil management technologies play a role in land-use changes. The changes, especially those advocating the use of fertilizer and manure, usually also lead to increased yields.

Introduction of improved breeds and seeds

In Arumeru, population increase has not only resulted in the deterioration of the agro-ecosystem, but also in the decline of the agro-ecosystems' capacity to support large livestock numbers. At present the majority of farmers in the high altitude zone cannot keep large numbers of cattle and goats despite the fact that milk, meat, and manure are still among the most needed resources for food and fertilization.

Exploitation of the biodiversity of natural forests

The biodiversity of natural forests adjacent to areas of land use is an important factor to local people's livelihoods. The forest provides many non-timber forest products and is also a source of plants for domestication on the farm. However, these forests, which are an integral part of the landscape, are under threat from overexploitation by people in search of timber and building materials. Other immediate threats include the migration of farmers from overpopulated areas to the fragile ecosystems in the forest buffer zones, poor farming practices, and the overexploitation of plants, animals, and other resources to meet their ever-increasing demands. According to 90 per cent of the *ilnyangusi* group (1930s), the forest cover on the slopes of Mount Meru was in earlier times thick with a rich diversity of tree and animal species. Most of these species are now rare or even extinct. The role of the natural forest in the agrodiversity of the whole landscape is thus being seriously challenged, and conservation needs to be targeted as part of an overall sustainable conservation effort.

Reintroduction of lost and endangered species

Farmers realize that there will be no recovery of the lost vegetation species unless deliberate efforts are made to restore the lost biodiversity. Farmer groups for tree-planting have been formed though PLEC initiatives. These farmers want to replace the tree species that were lost when the forests and scattered vegetation were invaded. To this end, PLEC farmers are planting lost species in nurseries, and the young trees are then distributed among other farmers.

Conclusion

This study has highlighted the significance of population pressure on land use and biodiversity in Arumeru. Other influential factors identified include changes in demand and the emergence of new opportunities through the advent of new field types and activities in the peri-urban interface. The major land-use and biodiversity change in Arumeru over recent years has been land fragmentation to accommodate family land and socio-economic requirements such as food, cash, and the need to utilize the soil and land resources better.

The majority of people in the study area are heavily dependent on agriculture and are likely to remain so in the near future. Land shortages, particularly in the densely populated upland regions of the area under study, mean that intensification of agriculture is a vital coping strategy with the simultaneous increase in population and decline in productivity.

This study has also emphasized the importance of tapping farmers' knowledge in understanding land-use and biodiversity changes. Therefore, to reduce the adverse effects of land and biodiversity degradation, the focal point should be farmers and their farms.

Another major lesson learnt was regarding farmers and their rational use of resources. Farmers are knowledgeable of the environment in which they live and are rational in the decisions they make and ways in which they use their resources.

There is a need to follow up, at short intervals, the major changes taking place in the area in order to generate information to assist efforts geared towards curbing land degradation and loss of productivity and biodiversity. A greater research effort is required to assess and evaluate land-use and biodiversity change and relate it to population growth in Arumeru.

Household food security and environmental sustainability in the Arumeru context also need further examination. Similarly, gross margin analysis of the various field types adopted at present and their relative contribution to farmers' socio-economic conditions and comparison of this with past field types could assist in advising farmers on crops to be cultivated given the available resources.

REFERENCES

Brookfield, H. and C. Padoch. 1994. "Appreciating agrodiversity: A look at the dynamics and diversity of indigenous farming systems", *Environment*, Vol. 36, No. 5, pp. 6–11 and 36–45.
Bureau of Statistics. 1988. *Population Census*. Dar ès Salam: Bureau of Statistics.

Morss, E. R. 1980. *Cross-cutting Issues Emerging from the Arusha Regional Planning Exercise.* Arusha: AP/VDP.

Padoch, C. and W. de Jong. 1992. "Diversity, variation, and change in ribereno agriculture", in K. H. Redford and C. Padoch (eds) *Conservation of Neotropical Forests: Working from Traditional Resource Use.* New York: Columbia, pp. 158–174.

Part III

Farmers' perspectives

14

Participatory technology development and dissemination: A methodology to capture the farmers' perspectives

Fidelis B. S. Kaihura

Introduction

Farmers are the custodians of knowledge and practices which scientists may use to develop better and more appropriate resource management technologies. There are many participatory approaches to technology development that involve farmers, including participatory learning and action (PLA), participatory technology development (PTD), farmer participatory research (FPR), and the participatory extension approach (PEA). Each has its own set of tools and methods, and the aims are as follows.

- Analyse community constraints and needs through participatory needs assessment (PNA) or participatory situation analysis (PSA). In these approaches communities are informers, providing baseline information from which scientists and external agencies may derive needs and circumstances.
- Joint identification of solutions and actions to overcome constraints through participatory rural appraisal (PRA) or participatory learning and action (PLA). In these approaches communities make their own responsible decisions, facilitated where required by external agents.

Differences between participatory approaches are often determined on how participation is applied.

- Passive participation. Communities are mere recipients of messages,

assistance and services. In this process communities have either to ac-
cept or reject suggestions.

- Active participation. Communities are consulted, providing information on constraints, needs, and even possible solutions. They carry out
  activities offered as solutions by service providers based on PRAs and
  PNAs. Communities have a choice, but final solutions are offered by
  outsiders.
- Interactive participation. Communities, either among themselves or
  jointly with service providers, interact in knowledge exchange, solution
  finding, decision taking, implementation, monitoring, and evaluation
  based on PRAs. Communities find their own solutions, make their own
  decisions, and take responsibility for results. They feel ownership for
  their actions and for subsequent results.

The PLEC approach in Tanzania (Figure 14.1) takes a further step to
empower farmers to train other farmers based on their own management
models. Successful farmers are given the title of "expert farmer" within

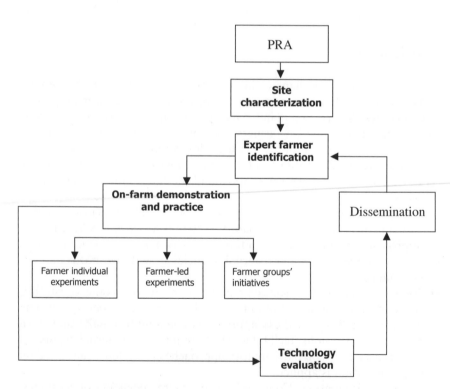

Figure 14.1 Participatory technology development and dissemination using expert farmers

PLEC. These expert farmers have accumulated knowledge and experience and have developed and modified coping strategies for the environmental changes they have experienced. Scientists from PLEC act as facilitators to integrate proven interventions with indigenous management models.

First, PRAs are conducted with farmers in order to identify constraints and appropriate management strategies that are environmentally, socially, and financially sustainable, and that protect biodiversity while also improving production and income. Unlike conventional mainstream agricultural research, PLEC integrates locally developed knowledge of soil, climate, biological resources, and other physical factors with scientific assessment of their quality in relation to crop production.

It is a systems approach carried out on farmers' own fields, taking current land-use types and cropping systems into account. Appropriate technologies are selected from scientifically proven options and integrated to improve the quality of farmers' own models without conducting any new experiments. A set of sustainable technologies is devised so that crop diversity and management diversity are maintained.

The process

Participatory rural appraisal

Discussions are carried out between farmers, extension staff, and researchers covering the following subject areas:
- key production constraints
- coping strategies
- extent of use of improved technologies
- evaluation of resource quality
- crops and cropping systems
- nutrient dynamics at farm level
- differences in farmer abilities to cope with climatic and biophysical changes.

Researchers and extension staff learn about farmers' resource assessment and management. Farmers are asked to draw maps of their farms indicating crops and cropping systems in each plot, reasons behind choice of crops and cropping systems, and intensity of management of each field type. Figure 14.2 pictures an older farmer (with a stick) educating the researcher (the present author) on plant indicators of fertile soils along a sloping land in Kiserian during a PRA exercise. The difference in this approach compared with standard approaches is that the farmer takes the lead where his or her knowledge is the greater.

Figure 14.2 A farmer training the researcher on diagnostic indicators of fertile soil in previously maize-cropped fallow

Identification of expert farmers

This process involves conducting close discussions with farmers considered particularly successful in resource management. Researchers and extension staff discuss the reasons for the successes and failures of certain management practices with the farmers. This also allows them to gain an understanding of the value of such management models within the cultural and social framework of the area. An evaluation of the farmers' communication skills, willingness to share experiences with other farmers, and the kind of respect they command amongst fellow farmers is made. Trainer expert farmers are then selected for specific management models. They are also asked to prepare teaching aids and sites for training, demonstration, and practice.

Expert farmers training other farmers

Expert farmers' successful management models are used as demonstration sites to teach other farmers. Management practices are demonstrated to other farmers and, where appropriate, researchers/extension workers provide additional scientific facts to support the performance of the practice and possibilities for improvement. Figure 14.3 shows one of the expert farmers in Ng'iresi village demonstrating agricultural intensification strategies.

In this particular case, the field is growing Irish potatoes that are nearly mature, while maize has been intercropped after weeding and ridging of potatoes. The practice ensures that there is a crop in the field throughout

Figure 14.3 Agricultural intensification demonstration in Ng'iresi village

the year. Behind the participating farmers is a narrow field planted with
the trees *Grevillea* and *Calliandra* along with yams – an agroforestry system to meet food, fodder, and timber household requirements from a
narrow boundary strip. The combination of a local expert speaking the
local vernacular and demonstrating practices that can be seen to be successful is powerful.

A demonstration site therefore takes on the role of classroom, with the
expert farmer acting as teacher while outside experts become facilitators
and participating farmers become both trainees and modifiers of the
technology. Interested farmers can take planting materials away with
them or learn about the availability of such materials, or can even exchange seeds and planting materials with the expert farmer. When the
training is recorded, the video is used to train farmers elsewhere or may
be used by participating farmers to revise what they have learnt.

Technology development is also carried out through farmer-managed
experiments. Through PRAs, resource management constraints are item-
ized and prioritized. Scientifically proven interventions are identified for
testing on volunteer farmers' fields. At crop development and maturity
stages, field-days are organized for farmers' evaluation of the performance of interventions and their potential for adoption. Figure 14.4 com-
pares maize performance according to soil fertility and water-harvesting
conditions with farmyard manure and tie-ridges (on the right of the
picture) with fertility improvement using manure only under semi-arid
farming conditions. During the field-day all farmers were pleased with
maize performance under manure and tie-ridging, but rejected the use of

Figure 14.4 Farmer-led experiment on fertility and water harvesting in semi-arid Kiserian village

tie-ridges and selected deep tillage and thorough manure incorporation (treatment not shown). Deep tillage was easy to adopt under cultivation with oxen and resolved the problem of subsurface hardpans created by ox-ploughing and consequent hampered water infiltration.

Training is also carried out at individual farmers' fields based on their own experiments (see Figure 14.4). Before PLEC, scientists did not know about the farmer experiments in Arumeru and the results were neither documented nor disseminated. Farmer experiments are diverse, dynamic, and private. They range from plant breeding, soil management, and pasture and forest management to pest and disease control. Few farmers discuss their research with anyone else or document their findings. Table 14.1 summarizes individual farmer experiments in Olgilai/Ng'iresi and Kiserian and shows how these initiatives contribute overall to agrodiversity.

Farmer experiments are partly in response to constantly changing environments or changes in resource potential. They are also a result of a lack of advisory services. Farmers also conduct experiments to evaluate the performance of techniques they learn about from outreach programmes such as farmer field-days, agricultural shows, and agricultural extension workshops. It is only through close interactions with such farmers that experts can recognize, discuss, and monitor their progress. Several of the commonly grown bean varieties in Arumeru are a result of farmers' own acquisition from friends, local markets, or their own crosses. They also obtain plant varieties from research institutions. Indi-

Table 14.1 Individual farmer experiments at Olgilai/Ng'iresi and Kiserian sites

Farmer	Experiment	Duration	Outcome	Implications for agrodiversity
Konyokyo	Maize/beans intercrop spacing	2 years	Spacing of 75 × 30 cm maize and 45 × 15 cm for beans changed to 90 × 30 cm maize and 30 × 15 cm for beans	Intercrop with legume beneficial for soil fertility; increased organic matter; greater below-ground biodiversity; better overall agrobiodiversity potential
Gidiel	Ratios of urine and water in making pesticides and accaricides for crop and animal pest control	3 years, ongoing	Urine/water ratio of 1:20 litres for vegetables and fruits and 1:20 plus some *Tithonia* spp. leaves once a week	Increased soil fertility and more organic matter; no artificial chemicals; good example of management diversity
	Sorghum, millet, and soya-bean production under sub-humid environments	First season, second season for soyabeans	Plant sorghum and millet in March/April and soya beans in June/July; sorghum and soya beans perform well in all soil types, unlike beans; millet performs well in fertile soils	Sustainable intensification concentrates production enabling adjacent areas to have greater diversity of use; example of organizational diversity
	Maize breeding of HB 622 vs *Larusa*	1997 and 1998	*Matatu* cross-breed with the following qualities: better milling quality; sweet, small cob with many grains; post-harvest pest tolerant; more tolerant to lodging when well spaced	Locally bred varieties more adapted to local needs; management diversity through employing local knowledge
	Use of different types of fertilizers for cauliflower production: ordinary soil, ashes, chicken manure, manure, forest soil; SA soda at planting and after second weeding	One season	Poorest performance with ordinary soil; premature leaf shedding with ashes; highest yield with chicken manure followed by farmyard manure; others comparable	Enhancement of soil fertility; good example of management diversity through employing locally available resources

165

Table 14.1 (cont.)

Farmer	Experiment	Duration	Outcome	Implications for agrodiversity
Navaya	Breeding for coffee tolerant to coffee berry disease	9 years	Good seedlings transplanted to the farm after 9 years; also source of proper seedlings proposed by neighbouring farmer during farmer field-day at his farm	Sustainable intensification of coffee and selection of natural biophysical diversity for production and income purposes
Logoro	Soil moisture conservation in semi-arid Kiserian	2 years	Ridging of maize and weed control to reduce soil moisture competition	Use of local knowledge in a marginal environment; management diversity
	Early blight control in beans in semi-arid environments	Continuous	Continuous selection of upright varieties leaving out others	Use of local knowledge in a marginal environment; management diversity
Kisioky	Evaluation of palatability of grasses in the conserved household pastures	Since 2000	Natural palatable grass species maintained (*Emurwai, Ologor-oing'ok, Olkujita-onyokie, Osangari, Enyoru*) and improved ones introduced (elephant grass)	Conservation of agrobio-diversity, especially of uncommon but nutritious grass species

vidual farmer experimental sites are alternative sites of technology development and demonstration. It is the project's intention to initiate joint research topics between farmers and researchers and improve on the quality and documentation of farmers' research outputs, while saving time and resources.

Training of farmer groups

Expert farmers are also influential in organizing farmer groups and group activities on resources management for biodiversity enhancement, production, and livelihood improvement. In most cases, expert farmers become leaders of such groups and influence group activities. Training is one major area of the group activities. Working with groups makes it possible to reach women and others who are often otherwise difficult to target in normal village visits. Women's groups have also been established with the assistance of female experts. Figure 14.5 illustrates a female expert training other women on local chicken-production alternatives – demonstrating different methods of raising chicks. Rearing chickens requires little initial capital outlay – the paper box costs US$0.6 or can sometimes be obtained free of charge. The improved box (in the background of Figure 14.5) costs US$15. Chickens are considered "women's cattle" and women have full responsibility for them.

Chickens also provide manure, improve family nutrition, and enhance biodiversity. After one year of training more than five women's groups

Figure 14.5 A female expert training a women's group on low-cost commercial production of local chickens

were formed and more than 70 per cent of households kept chickens. These groups are dynamic units and are considered a useful method of information sharing, technology development, and dissemination among women, but also reaching neighbouring and distant interested villages.

Active participation of farmers in stakeholder workshops

Another tool for participatory technology development and dissemination is participation of farmers in stakeholder workshops. Figure 14.6 indicates one of the poorest widows of the area (with only 0.12 ha of land) explaining to workshop participants how she is diversifying production on her small piece of land.

Crop/livestock interaction interventions have helped improve her food requirements and provide her with some cash from egg sales. Such poor farmers – particularly women – are often left out in participatory rural development activities. Poor farmers are often the most important custodians of biodiversity because of their reliance on the sustainability of local natural resources.

Workshops have also been used for training policy-makers and decision-makers who do not frequently visit farmers in their villages. From these workshops, policy-makers help sensitize other farmers and experts to undertake and/or demonstrate good practices. Decisions are

Figure 14.6 A female farmer addressing delegates at a PLEC stakeholder workshop

often made during or after farmers, experts, and policy/decision-makers engage in interactive deliberations in PLEC workshops and meetings. Previously it was not common practice to involve farmers in meetings or workshops that deliberate on resource management practices or small-scale farming and rural development.

Advantages of the methodology

Farmers indicated that during different workshops, technologies developed address existing constraints and affect their daily life. Experiments and demonstrations are owned and managed by well-known and respected farmers.

The adoption of new technology with or without modification is based on the farmer's own assessment and is advocated and promoted by farmers themselves instead of extension and research staff. The process starts from farmers' own existing successful management practices. These practices may be modified from scientifically proven technologies. Adoption is gradual, and is continually evaluated on the farmers' own fields. Adoption may also take time depending on the types of farmers in the partnership with researchers and extension agents.

Dissemination is rapid and extensive. It starts immediately with participants at the demonstrations and, as those participants share the information with friends and family, it can spread to other villages, districts, regions, and ultimately the whole nation.

Improved technologies depend on the integration of farmers' tried-and-tested techniques along with scientific assistance where appropriate. Instead of development and dissemination taking long periods of experimentation and evaluation, new and more appropriate technologies arise far more quickly and are disseminated more effectively. The process is both cost-effective and practical. However, the process is intensive in terms of the time and availability of researchers because of the need for frequent visits and continuous interaction with farmers.

Conclusion

The PLEC-Tanzania approach of PTDD is centred on expert farmers as trainers of other farmers. Successful models of resource management are demonstrated to other farmers. The training provides the opportunity for exchange of knowledge and experience as well as materials, testing and/or modifying techniques to fit individual farmers' field conditions. Experts facilitate the process by introducing ways of integrating scientifically proven interventions into local models to improve their quality or effectiveness. School children who are future custodians of the resources are also currently being involved in on-site farmer training sessions.

The methodology requires experts to work as equal partners with farmers in information sharing and implementation. As well as demonstrations of successful models on the expert farmers' fields, there are other tools that facilitate the process of information exchange, technology development and testing, and the adoption and dissemination of new technology with the farmer in the steering seat. These include the use of individual farmer experimental sites; joint researcher/farmer/extension worker research projects; farmer field-days; and farmers' participation in stakeholder workshops. Through these methods, technology development and dissemination to many farmers is fast. However, further testing of the approach is recommended before its wide adoption.

15

Experimenting with agrobiodiversity conservation technologies: My experiences with *setaria* grass in a banana plantation in Uganda

Frank Muhwezi

Introduction

I am Muhwezi Frank Kashaija, a PLEC demonstration farmer and teacher at Kikunda primary school in Bushwere parish, Mwizi subcounty, Mbarara district. I have also been a member of the Mwizi PLEC Experimenting Farmers' Association (MPEFA) since 1999 and am a member of the Bushwere Zero-grazing Crop Integration Association (BUZECIA) which we formed in 2000.

I bought a banana plantation in 1984 – it had a lot of couch grass (*orumbugu*), wandering Jew (*eteija*), and local Irish potatoes (*emonde*) as weeds. So I removed them, for they were competing with crops for moisture and nutrients. The banana plantation had a poor performance so I rehabilitated it. I observed some loss of soil by water erosion due to the slope of the land from north-east to south-west. So with help from our agricultural officer, Mr Mpiirwe, I dug trenches to check soil erosion and harvest water to increase water penetration into the soil. I copied this idea from Mr Kaakare, who had a worse plantation than mine before he constructed trenches with *setaria* (a type of grass), but later his bananas improved.

Efforts to control soil erosion and harvest rainwater in the plantations of the Bushwere Regional Land Management (RELMA) project included soil and water conservation structures – trenches planted with *setaria* grass (*fanya chini*). Being a science teacher, I knew that in theory grasses could help to reduce soil erosion by their root-binding action and formation of humus after decay. So I planted *setaria verticillata* (*orutaratumbe rwa* Tanzania) to stabilize the bunds (earth ridges along the contour) of my trenches (which I found useful in controlling erosion), and my banana production started to improve.

Later on, I heard a rumour that *setaria* grass bunds consume much water, thus causing moisture shortage in plantation soil. Most farmers in Bushwere and elsewhere in Mwizi uprooted it from their banana plantations. Having some knowledge of science, I refused to uproot the *setaria* without proof because I did not have a better alternative to trap the runoff and stop erosion. So I set up my locally designed experiments.

I widened the spacing of the *setaria* to about 6 ft,[1] which is 5 ft more than the 1 ft intervals recommended by the agricultural officers, and left one plot aside without *setaria*. In due course PLEC arrived in Bushwere. As they were carrying out their research, I told them the *setaria* rumour and how I would like to experiment on it. They welcomed my idea and encouraged me to work with them.

My study with PLEC

An experiment was set up with the following objectives:
- to evaluate the effectiveness of different spacing of *setaria* grass in controlling runoff soil loss
- to evaluate the allegation that the *setaria* grass absorbs more water, causing shortage of soil moisture in the plantation
- to compare the profitability of different spacing of *setaria* on banana and *setaria* production.

I persuaded two other farmers to set up participatory field experiments in their fields with plots using close spacing (1 ft apart), i.e. the spacing recommended by extension workers, distant spacing (6 ft apart), and one without grass (a control).

The extension workers also advised me to plant the *setaria* grass on the upper side of trenches (*fanya juu*), while the PLEC scientists taught me how to manage the *setaria* grass bund. This was by trimming it at the right height before flowering, maintaining its width at about 1 ft, and keeping the grass at a 1 ft distance from the banana stools. They encouraged me always to throw the soil I clean from the trench on the upper side (*fanya juu*) beyond the grass bund.

I had started harvesting the *setaria* for my goats and also mulching the bananas. During the dry season my friend Kaakare would take it for his cows, and I could see that soon there might be a market for it among zero-grazing farmers so I could earn some cash income from it.

We agreed to take the following measurements:
- soil deposits from soil erosion in the trenches
- soil moisture levels in three parts of each treatment
- weight of banana bunches harvested from each treatment
- labour costs of maintaining the grasses and trenches
- the amount of grass I harvested each time I trim it (to estimate the income that I would get if I was to sell it or feed it to livestock)
- weight of beans harvested from each treatment – I sometimes intercrop beans in the banana plantation for food security reasons.

The PLEC scientists helped me to do the technical measurements and analysis of soil loss and soil moisture, which I could not manage to do on my own. The soil deposited in the trenches from the runoff was determined using methods described by Stocking and Murnaghan (2001). Soil loss from the plots between the trenches for each treatment was calculated (in tonnes per ha) from the soil deposits. The data were then statistically analysed to determine the effect of different spacing of *setaria* grass bunds on controlling the run-off.

The soil moisture levels in the plots of each treatment were monitored by collecting samples during the dry season at two depths: one from the topsoil (between 10 and 15 cm), the rooting zone for *setaria* grass, and the other from the subsoil (between 30 and 35 cm), the rooting zone for banana.

For each treatment, samples were taken from three locations:
- banana sampling location: at least 30 cm away from banana stools and between neighbouring trenches
- *setaria* sampling position: 30 cm above the *setaria* bund
- trench sampling position: 30 cm below the trench.

Soil moisture

The data for percentages of available soil moisture are presented in Figure 15.1.

The results show that the available soil moisture in the banana sampling position was significantly higher for the closed *setaria* treatment than the spaced *setaria*. A similarly significant situation was noted in the *setaria* sampling position, which shows that a closed *setaria* treatment conserved more soil moisture than an open-spaced one. The differences between closed *setaria* treatments and the control were not statistically significant, but observations suggest that available soil moisture was higher in the closed plot.

However, it should be noted that closed *setaria* treatment had other benefits, as shown in Figure 15.2 and Table 15.1.

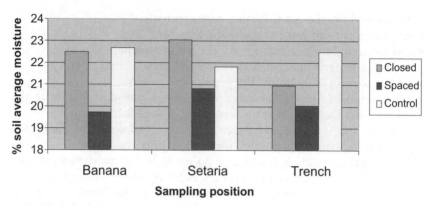

Figure 15.1 Variation of percentage soil average moisture with sampling position and treatment

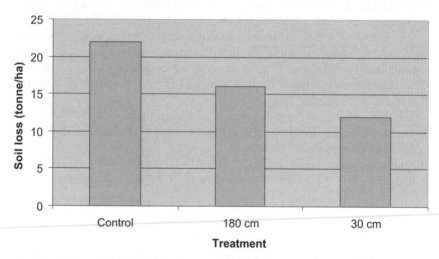

Figure 15.2 The effect of spacing *setaria* grass (within an SWC bund) on soil loss in banana fields

Table 15.1 Average soil loss and relative soil moisture content per season

Treatment	Soil lost (tonne/ha)	Moisture content	Labour requirement to clean trenches	Water in trenches after heavy shower
Closed grass (1 ft)	12	Higher	Lowest	Clean water
Distant grass (6 ft)	16	Moderate	Moderate	Dirty water
No grass (control)	22	Less (little)	Highest	Lots of mud

Therefore, the recommended spacing of *setaria* grass (1 ft) was significantly ($p = 0.05$) more effective in controlling soil runoff in banana fields than the farmers' spacing (180 cm) or the control.

Expenditure and income

Profit from *setaria* grass was also found to be significantly higher when closed spacing was adopted. Table 15.2 summarizes the expenditure incurred and the predicted income of planting *setaria* grass.

With regard to income and expenditure, it should be noted that:
- in the wet season, trimming and cleaning are carried out about four times in a three- to four-month period, depending on climatic conditions
- in the dry season, trimming is done twice in three months, and cleaning of trenches is done once when rains are expected
- the price of beans is Ush200 per kilo.

Advantages of planting *setaria* grass

There are seven other farmers in Bushwere who became interested and started their own experiments on *setaria* grass management in banana plantations. From my experience, I found *setaria* grass to be useful to farmers for its role in reducing soil erosion. Similarly, it can be used as mulch and therefore to conserve moisture in the soil and to suppress weeds.

It can also be used as animal feed for goats, sheep, and cattle (as done by Mr Kaakare and Fred Tuhimbisibwe). Those animals then produce manure which can be used to replace the lost plant nutrients (recycling of nutrients) in plantations and increase banana yields. Scientists explained to me that there is a high protein content in *setaria* grass tissues (about 2.5 per cent nitrogen), which means that it could increase milk yields when fed to livestock. Another farmer who I visited (Mr Gita) confirmed this.

Growing *setaria* grass in plantations increases the biodiversity (i.e. it is another way of conserving biodiversity). Finally, all of the suggested benefits play a role in increasing income and family welfare.

Conclusion

In conclusion, I am convinced that *setaria* grass is good for controlling soil and water losses. It does not absorb all the water if well managed.

Table 15.2 Expenditure and income from each plot

| Plot | Time taken (min) | | Cost (Ush) | | Total cost/plot (Ush) | Income from *setaria* (Ush) | Income from banana (Ush) | Income from beans (Ush) | Total income per plot (Ush) | Profit (Ush) |
	Trimming	Cleaning	Trimming	Cleaning						
A	60	40	500	200	700	500	3,200	5,200	8,900	8,200
B	45	60	300	300	600	300	1,950	5,000	7,250	6,650
C	–	90	–	500	500	–	1,350	4,600	5,950	5,450
Total	**105**	**190**	**800**	**1,000**	**1,800**	**800**	**6,500**	**14,800**	**22,100**	**20,300**

Since it has many other uses, I recommend that other farmers should keep it in their plantations, but they must learn to manage it well by planting it on the upper side and keeping it well trimmed and not too close to the banana stools. I also advise my fellow farmers to replace the useless and stubborn weeds in gardens with useful biodiversity, as I replaced couch grass and other weeds with *setaria* grass bunds.

Now that I have learnt the many good reasons to plant *setaria* grass and how to manage it well, I am planning to start a zero-grazing cattle unit. This will also encourage me to conserve more useful biodiversity like *Calliandra*, *Desmodium*, and Guatemala grass on my limited agricultural land and around my homestead.

I would like to thank PLEC project staff who have tirelessly helped us farmers of Bushwere to test some of the agrobiodiversity technologies. It is helping us to understand our farming in a more scientific way. We now know why we should carry out certain farming practices if we are to have high yields. Also I would like to thank PLEC who have facilitated me to come and attend this workshop to share our research results with you. We invite you to Bushwere parish, Uganda, so that we can demonstrate to you good practices of conserving biodiversity in crop fields. Remember that PLEC has developed Bushwere into a demonstration site, so come and use it for training other farmers and doing more research.

Finally, I thank you for listening.

Thank you. All this, for God and my country, Uganda.

Note

1. Throughout this part of the book, imperial as well as metric measurements have been cited, as these are the measurements used by the participating farmers.

REFERENCE

Stocking, M. A. and N. Murnaghan. 2001. *Handbook for the Field Assessment of Land Degradation*. London: Earthscan.

16

The shared activities of PLEC farmers in Arumeru, Tanzania

Frida P. Kipuyo, Gidiel L. Loivoi, and Kisyoki Sambweti

Introduction

Work with PLEC started in 1998 at sub-humid Ng'iresi/Olgilai and semi-arid Kiserian. It started with site selection and soils' characterization, followed by on-farm experiments. Different inputs were given to farmers in Kiserian and Ng'iresi/Olgilai. Initial experiments were unsatisfactory in Kiserian (due to bad weather) and average in Ng'iresi/Olgilai.

The following year (1999–2000) a few farmer groups were started. These covered local chicken production and environmental conservation. More farmer groups were initiated later in the year and earlier ones strengthened. By May 2001, 11 farmer groups had been established.

Through the project it has been possible for most farmers to visit research institutions such as Selian in Arusha, Lyamungo in Kilimanjaro, and Ukiriguru near Mwanza, where they received training.

Outcomes

With increased production and income from food and cash crops, services were brought closer to farmers. Other items previously not available in the villages became readily available. Examples include tree seedlings, improved pastures, eggs, milk, vegetables, chicken, rabbits, and guinea pigs. Other activities included planting indigenous trees in water sources and rivers, farmer exchanges of knowledge and practices, identification and conservation of indigenous endangered trees and other plant species, adding value to traditional foods, and sharing indigenous knowledge of traditional medicine.

Researchers, extension workers, and other experts valued the farmer activities highly. There was full-time close collaboration between farmers, researchers, and extension workers in all crop and livestock production activities. Farmers were also trained in accounting for income and expenditure at household level.

Through PLEC it became possible for some of the farmers to obtain improved breeds of cattle, pigs, chickens, dairy goats, and sheep. Agricultural inputs were also made more readily available to farmers. Expert farmers were rewarded with different farm implements like wheelbarrows, machetes, spades, and irrigation pipes.

Conclusion

The collaborative work of farmers (with assistance from PLEC) has therefore been successful in the case of Arumeru, Tanzania, and the farmers involved have high expectations for their future. To continue and strengthen knowledge obtained through the PLEC project and to spread this knowledge to neighbouring villages, regions, and the whole nation, training for PLEC farmers should continue in order to meet national targets of food sufficiency and improved livelihood.

Editors' note

Comments for this paper were originally delivered verbally in KiSwahili. The original Swahili is retained below to capture the farmers' enthusiasm and to enable them to read at least a small part of this book.

Mwaka 1998 ndiyo mwanzo wa mradi huu katika vijiji vya Ng'iresi/Olgilai katika ukanda wa juu wa Kiserian na ukanda wa chini.

Walianza kwa kuchagua maeneo ya kufanyia kazi na pia wakachukua udongo tayari kwa kuupima.

Baada ya udongo kupimwa na matokeo kujulikana wakulima walipewa mbegu, mbolea kwa ajili ya utafiti wa maeneo hayo ya Kiseriani na Olgilai/Ng'iresi. Kwa ukanda wa juu walitoa mbolea pekee kwa ajili ya mahindi na viazi mviringo.

Matokeo ya utafiti wa mwanzo

Kiseriani: kwa ujumla matokeo hayakuridhisha kwa sababu ya hali ya hewa.

Olgilai/Ng'iresi: matokeo yalikuwa ya wastani.

Msimu uliofuata wa mwaka 1999–2000 vikundi vichache viliundwa vya ufugaji wa kuku wa kienyeji na uhifadhi mazingira. Baada ya kupita miezi mitano vikundi vingine vilianzishwa na huku vile vya mwanzo vikiimarishwa. Hadi kufikia mwezi May 2001 vikundi 11 vimeisha undwa pia:

- vikundi vya mazingira viko (5) ukanda wa chini
- vikundi vya kuku viko (2) viwili ukanda wa chini
- vikundi vya mazingira viko (2) viwili ukanda wa juu
- vikundi vya mazingira (shule) viko (2) viwili chini na juu
- kuna baadhi ya vikundi binafsi (kaya).

Mafunzo/ushauri

Mradi huu umetuwezesha sisi wakulima kuungana na kupeana mafunzo na mbinu mbalimbali hasa juu ya shughuli zote tunazofanya katika vijiji vyetu na vya nje.

Ziara ya mafunzo

Mradi umetuwezesha kwenda kutembelea na kuona au kupata mafunzo mbalimbali kwenye vituo vya utafiti vilivyoko ndani na nje ya mkoa. Mfano: Chuo cha utafiti Selian-Arusha, Chuo cha utafiti Lyamungo-Kilimanjaro na Kituo cha utafiti cha Ukiriguru-Mwanza.

Mafanikio

Kipato kimeongezeka.

Uzalishaji umeongezeka wa mazao ya chakula na biashara.

Huduma kuwa karibu.

Mfano: miche ya miti, majani, mayai, maziwa, mboga za majani, nyama ya kuku, sungura, sili na k.n.

Miti ya asili kwenye vyanzo vya maji au mito.

Kubadilishana utaalamu wakulima kwa wakulima na kuelimishwa kutokana na mafanikio binafsi ya wakulima mashambani mwao.

Kutambua na kudumisha miti ya asili, vyakula, madawa na Elimu ya asili.
Watafiti kuthamini shughuli za wakulima.
Wakulima, watafiti na wataalamu wa ugani kuwa karibu katika shuhguli za kilimo na ufugaji.
Kutunza kumbukumbu za mapato na matumizi ya kila siku.

Mradi wa PLEC umetusaidia kwa ushauri jinsi ya kupata mifugo.

Mradi umetusaidia katika kuchangia upatikanaji wa mifugo kama ng'ombe, nguruwe, kuku, mbuzi wa maziwa, kondoo n.k.
Mradi wa PLEC pia umesaidia katika kuchangia upatikanaji wa pembejeo kama vile mbolea na zana za kilimo.
Mradi pia umetufundisha juu ya kutunza kumbukumbu za mapato na matumizi ya kila siku.

Wakulima wengine kupatiwa mifugo mfano: nguruwe, mbuzi wa maziwa, kondoo, kuku na ng'ombe.
Wakulima wengine wamebahatika kupewa whee-barrow, panga, jembe beleshi seng'eng'e ya kutengenezea wigo na wengine kupatiwa plau ya kutengenezea makingo.
Wakulima wengine wamepewa vyakula vya kulishia mifugo kama: pumba; kwa ajili ya kuku na ng'ombe.
Wengine walipewa kilo za kupimia uzito wa mazao yao.
Kuna wakulima waliopatiwa bomba kwa ajili ya kuwaogesha mifugo, kunyunyizia mboga dawa na kahawa.

Mategemeo yetu wakulima katika mradi wa PLEC

Kuendelea na kudumisha Elimu au ujenzi, tuliopata kwa kuuagawia wakulima majirani na jamii yote kuanzia ngazi ya kaya hadi Taifa, yaani: familia zetu, majirani, vijiji, kata, tarafa, wilaya hadi taifa zima.
Tunomba mradi wa PLEC upanuke kwa maeneo ya kazi.
Wakulima waanzilishi waendelezwe kwa kupewa Elimu zaidi – ili waweze kufikia malengo ya Taifa.

17

My experience as a farmer with the KARI PLEC team

Bernard Njeru Njiru

Introduction

I am a farmer from Nduuri village, Kagaari North location, Embu district in the eastern province of Kenya. I am married to one wife and have four children – two boys and two girls. I am a mixed subsistence farmer with five acres [approximately two hectares] of land situated on the slopes of Kirimiri Hill, near Runyenjes municipal council town. I have about 1,000 grown-up coffee stems on the farm, plus some mango trees, avocados, and macadamia trees. I also grow food crops like maize, beans, yams, bananas, and sweet potatoes among others. I keep about four dairy cows and a few dairy goats of the *torgen burg* breed.

Working with PLEC

The PLEC project team found me in the middle of grafting my old coffee stems with *Ruiru* 11 variety. The reasons for doing this were as follows:
- to avoid loss of coffee berries during cold weather (due to coffee berry disease)
- to avoid loss of coffee leaves during epidemics of leaf rust, which occur twice a year during dry seasons
- to reduce the high cost of fighting the above fungal diseases

- to avoid the expense of uprooting the old coffee stems in the farm and digging new holes to plant *Ruiru* seedlings
- to avoid the Ksh15 per plant cost of buying *Ruiru* 11 seedlings
- to minimize the amount of time that it takes for new plants to mature and produce effectively (usually three to four years for new plants).

Therefore, I had decided to take my time and do the grafting. The KARI (Kenya Agricultural Research Institute) PLEC team found me working on the above and I thank them for encouraging me to continue and for providing me with polythene tubes (for seedlings and cuttings), which have helped me in the work. Currently I have about 300 productive grafted stems among the 500 stems I had worked on, and I still have more to do.

Nduuri community, like other Kenyan communities, has an over-reliance on foreign vegetables, food crops, and cash crops like coffee and wood trees. This is as a result of assimilation with the European culture and way of living. When we compare ourselves with our ancestors, who were healthier, stronger, and well-nourished persons, we find that it was because of the kind of foods they used to eat. These included herbal medicines from the collection of the vegetables, root tubers, and fruits. This kept them and their livestock free from most tropical diseases. It is unfortunate that most of this local indigenous vegetation has slowly disappeared from our environments.

Since the introduction of the PLEC project in Nduuri in 1998, farmers in the area have learnt a lot from KARI through informal education on the need to preserve our indigenous food crops, especially root tubers like yams, sweet potatoes, cassava, vegetables, and bananas. They have also learnt the importance of using animal and compost-pit manure to improve soil texture – long gone through overuse and soil erosion in our smallholder farms. They have also emphasized the importance of terracing and planting grass on the terraces to prevent soil erosion.

The approach used by KARI PLEC involves learning from farmers about the usual way of farming. Reasoning and working together has been very encouraging to the farmers. It has created a friendly atmosphere because the farmers were able to learn through seeing, touching, doing, and where possible tasting, and also through visiting other local and distant farmers. This has greatly helped farmers to understand where they have been going wrong, where they are right, and how to improve and increase production and income.

I personally have learnt from other farmers that growing *miraa* can be a viable farming business within my area. At Kigumo in Kyeni location, Embu district, we visited a farmer with very steep and barren land who, after experiencing many difficulties, had discovered that mango trees could do well there. He planted about 1,000 mango trees, which earn him hundreds of thousands of shillings during the mango high-peak season. Today he is very comfortable with his family in his good stone house. He told us that to avoid soil erosion in this steep farm, he had to dig terraces. He only slashes the growing bushes to keep his mango plantation clean.

We visited another farmer in Chuka (Meru South district) with a small piece of land who had learnt to dig a borehole around his homestead. Today he is managing a big nursery farm of flowers and fruit seedlings where he is the managing director and sales manager and one of his sons is the farm manager. They sell seedlings both within the local town market and outside. The family seems to be very settled by skilfully managing the farm and generating a good income from their own initiative.

Conclusion

Today, great change can be realized in our farm management because we have learnt a lot from other farmers. Similarly, taking care of the natural vegetation within our environment is important, since it is directly or indirectly beneficial to us as farmers in Nduuri. Through the PLEC team, farmers in Nduuri have been able to identify our strengths, weaknesses, opportunities, and threats in farming through group discussions and farmer-to-farmer visits. In this spirit, farmers are now able to eradicate poverty and be self-sufficient in both market and domestic food. It is our sincere prayer as farmers in this area (Nduuri) that the PLEC project be allowed to continue, because we feel there is much more we need to learn from them.

18

Agrobiodiversity conservation for the promotion of apiculture and household welfare in Uganda

Fred Tuhimbisibwe

Introduction

I started farming in 1987 when I inherited fallow land that was used to grow annual crops from my father. I started growing beans and sweet potatoes before planting bananas. I then invested my retirement package (from the UPDF in 1992) in crop production and beekeeping. I planted about one acre [0.4 hectares] of bananas in which I conserve a lot of biodiversity near my apiary (as indicated in Table 18.1). Some plant species like avocados, local tomatoes, mango, guavas, and some medicinal plants were volunteer plants in the banana garden, but I deliberately conserved them because they are useful. Then I planted species like passion fruit, eggplant, modern avocados, *robusta* and *arabica* coffee, *Calliandra caloyrthus*, *Leuacena* spp., castor oil, pigeon peas, *setaria* grass, velvet beans, and pumpkins. Around my apiary I also have other fields grown with eucalyptus, cassava, sweet potatoes, cyprus trees, flame trees (*Erythrina abyssinica*), *Ficus* spp., Irish potatoes, maize, sorghum, and elephant grass. All these plant species give pollen and nectar to the bees and some are used for feeding my cow and goats. At the edge of the plantation, I planted some trees like eucalyptus and patula (*Pinus patula*) to stop the wind from destroying my bananas. The bees use the sap from some waxy trees like patula and wild cassava to repair their hives. I use the flame trees and wild cassava to support beehives.

Table 18.1 Benefits of my collaboration with PLEC

Before collaboration	PLEC intervention	Current situation
Adverse effects on the banana plants during dry seasons Also, poor performance of bananas due to high populations of many different plant species intercropped, and improper spatial arrangement of the different plants in the field	Training in better management techniques for different plant species in the banana plantation, e.g. planting *setaria* grass along soil/water conservation structures and both *setaria* and elephant grass on boundaries; also, the importance of correct spacing between plants	Banana plantations well managed throughout the year Increased yields Increased fodder for livestock from harvested *setaria* grass
My orange and avocado trees could not yield well	Facilitated me to acquire high-yielding, grafted orange and avocado fruit trees Taught me to manure well	I am expecting increased fruit yields and increased income through fruit sales
No farm record-keeping and no concern for costs or benefits	Trained me in record-keeping for better farm management	Improved record-keeping and able to do cost/benefit analysis for my farm
High post-harvest losses of maize, beans, and Irish potatoes Low prices during harvest	Trained me in modern storage facilities for farm produce	One improved store for maize and Irish potatoes I am now able to sell my produce when prices are high
I used to grow whatever I wanted, even interplanting in bananas without consulting my wife	Sensitized us to gender issues and enabled me to be gender-sensitive in farm operations	Harmony now prevails in my family as the workload is now shared amongst members
Failing to cooperate with my fellow farmers	Facilitated me to work and plan with other farmers by formation of associations	I can work easily with other farmers in Bushwere and I am now an active member of BUZECIA; we have now started writing up proposals for funding; I am also chairman of MPEFA

Table 18.1 (cont.)

Before collaboration	PLEC intervention	Current situation
I had fears of owning a good breed of cow because of lacking good fodder I used to lose my bees as they went a long way in search of pollen and nectar	Encouraged and enabled me to integrate and manage properly different plant species on my farm for livestock and beekeeping	My farm has many flowering plants from which bees collect pollen and nectar My bees are always busy in my fields and their hives, therefore I harvest more and better honey; each jellican of honey is worth Ush60,000 I bought a cross-breed cow, which I call Mercy Better fodder for my cow and goats; I am expecting a lot of milk from the variety of fodder species I feed to the cow
I lacked ability and self-confidence to share experiences with others, and did not encourage my wife to attend meetings and training sessions	PLEC built my capacity and confidence by helping me to see that I am a good farmer and explaining to me the good and bad impact of my practices; I have learnt a lot from farmer exchange visits organized by PLEC, and they have encouraged us to go with our wives to meetings, workshops, and demonstrations	Improved self-esteem and communication skills Increased experience, confidence, and organizational skills; I have given demonstrations to three farmer groups, two workshops, and to many visitors and local leaders; my wife attends workshops etc. and is also a member of BUZECIA; she is happy and willing to try out and manage well new practices on the farm

Benefits of working with PLEC

I used to have very little help with my farming, but since 1998 my family has been collaborating with PLEC scientists on land management and agrobiodiversity conservation. Table 18.1 summarizes the benefits I have experienced from this collaboration.

There have been many improvements to my farm and farming practices, including about 20 modern hives. I have planted more trees in my apiary, which are good sources of nectar and pollen for my bees. I continue to learn by attending training sessions in beekeeping technology.

By increasing the area of land I cultivate and by planting many species I have contributed to an increase in biodiversity. This has helped to meet the feeding needs of my cow Mercy, which calved and gave me a heifer which I named Gift. My family now has milk to drink and daily income from milk sales. My neighbours buy milk from me.

I have sensitized other farmers' groups like BUZECIA, the Kirinju (elderly) group of Bushwere, BUDEG, and the Karuhiira group whose members have started beekeeping, and I am now a committee member of Ankole District (Mbarara, Bushenyi, and Ntungamo districts) Bee Keeping Association.

Conclusion

On behalf of my family and myself, I would like to thank the PLEC scientists for increasing my capacity to conserve and use the natural resources I have in a sustainable way. Also, I thank the organizers of this workshop who have facilitated me to come and share my experience with you. I also thank you for listening to me. Finally, I invite you to come and visit my farm in Bushwere demonstration site, in Mwizi subcounty, Mbarara district.

19

Farmers' evaluation of soil management practices used in Mbarara district, south-western Uganda

Jovia M. Nuwagaba-Manzi and Joy K. Tumuhairwe

Introduction

A subsistence farmer's ability to maintain agricultural production depends heavily on his/her efforts towards maintaining soil quality through proper management of the soil resource. In Uganda, concerns over correct soil management started long before the colonial government. Soil conservation is a major feature of land resource management in south-western Uganda (ICRAF 1997). The area is characterized by hilly terrain and a rapidly growing population. It is therefore prone to soil degradation as a result of over-cultivation and accelerated soil erosion (Tumuhairwe *et al.* 1999). Various stakeholders have intervened in promoting soil management practices in order to combat soil degradation. Consequently, farmers in this area have been using both traditional methods and other soil management techniques introduced by government extension staff and non-governmental organizations (NGOs).

Farmers' perceptions determine their utilization of different soil management techniques. According to Brookfield (1989), farmers' land management decisions are not separate from decisions of production, consumption, and social control. Farmers use various frames of reference to appraise the relevance and usefulness of research and development products accessible to them. When appraising interventions from various sources, farmers consider the expected added value in respect to their objectives, the practicality of the proposed intervention, and its fit within their ongoing practices (Opio-Odongo 1999). Therefore, effective research and understanding of decisions at farm level (especially how farmers manage natural resources) is essential in this type of research (Loevinsohn and Wangati 1993).

Consideration of the decision-making process that farmers use when choosing which interventions to implement is crucial in promoting relevant soil management practices that aim to achieve sustainable production. Involving farmers in evaluation of the various practices would help researchers to develop acceptable and affordable technical options and suggest more effective extension approaches (Loevinsohn and Wangati 1993).

This study was part of a broader study on characterization and evaluation of agrodiversity and agricultural biodiversity in the UNU-PLEC collaborative project. A survey was conducted in Mwizi and Kabingo subcounties in the Mbarara district in south-western Uganda. The target population was 6,028 farming households in Bushwere and Kamuri parishes where the PLEC project was developing demonstration sites for promoting sustainable approaches to land management and the conservation of agricultural biodiversity. A total of 20 villages were covered. Using a systematic random sampling technique, 120 households were selected. This involved selecting one from every eight to 10 households (depending on the number of households) in each village. Data were collected through semi-structured interviews.

Major traditional and introduced soil management practices

Diversity of soil management practices was evident in both Mwizi and Kabingo subcounties. Over 15 traditional and five introduced practices were reported, but in order to carry out an in-depth analysis a manageable number of practices were selected. Arbitrary percentages as cut-off points were used to determine the major traditional and introduced soil management practices. The major traditional practices were defined as those that were used by 70 per cent or more of the total respondents, and the major introduced practices as those utilized by 20 per cent or more of total respondents. The major traditional and introduced soil management practices are presented in Table 19.1.

Mulching and utilization of household waste were the most commonly used traditional practices (over 90 per cent), because all farmers in both Mwizi and Kabingo subcounties grow bananas where these practices are mainly used and in most cases the banana plantations are located near the homesteads. The two communities also eat bananas as their staple food (Tumuhairwe et al. 1999). Banana skins along with other house-

Table 19.1 Major soil management practices (n = 120)

Traditional practices	Frequency	%	Introduced practices	Frequency	%
Mulching	110	91.7	Trenches	57	47.5
Household waste utilization	109	90.8	Controlled bush and trash burning	44	36.7
Crop rotation	92	76.7	Soak pits	27	22.5
Intercropping	90	75.0			
Rough tillage	110	91.7			
Fine tillage	96	80.0			

hold waste are routinely applied to banana plantations as mulch and to improve soil fertility. The remaining major traditional practices are in annual crop fields. The only major introduced practices were trenches, controlled burning (bush and trash), and soak pits, used by 57, 44, and 27 respondents respectively. Introduced practices were not widely used, probably because of factors associated with low adoption rates (discussed in Busingye, Tumuhairwe, and Nsubuga 1999). Trenches and soak pits are used only with banana cultivation, which requires more water than cultivating other crops. Trenches and soak pits are therefore primarily constructed for rainwater harvesting and soil erosion control.

Reasons for using major traditional soil management practices

Farmers' evaluations of major soil management practices were based on the perceived advantages and disadvantages of the different practices. Table 19.2 presents farmers' reasons for using major traditional soil management practices, while Tables 19.3 and 19.4 present farmers' perceived advantages and disadvantages of the traditional practices.

Mulching

The majority of farmers used mulching because of its effect on weed control (59.1 per cent), while others recognized its role in soil fertility improvement (28.2 per cent) and water conservation (10 per cent). These reasons were perceived as advantages of mulching, as were its low labour requirements (15.5 per cent) and inexpensiveness (17.3 per cent). The majority of farmers' perception of weed control (56.4 per cent) being one of the advantages for mulching implied that use of mulch avoided the costs incurred in weeding and left more time for other activities.

Table 19.2 Farmers' reasons for using the major traditional soil management practices

Reasons for utilization	Mulching		Household waste utilization		Crop rotation		Intercropping		Rough tillage		Fine tilth	
	F	%	F	%	F	%	F	%	F	%	F	%
Weed control	65	59.1	27	24.8	–	–	–	–	–	–	–	–
Conserve water	11	10.0	–	–	–	–	–	–	–	–	–	–
Soil fertility improvement	31	28.2	63	57.8	43	46.7	–	–	13	11.8	–	–
Get good yields	–	–	–	–	49	53.3	–	–	9	8.2	10	10.4
Food security	–	–	–	–	–	–	76	84.4	–	–	–	–
Inadequate land	–	–	–	–	–	–	22	24.4	–	–	–	–
Opening virgin land	–	–	–	–	–	–	–	–	47	42.7	–	–
Aid good germination	–	–	–	–	–	–	–	–	–	–	43	44.8

Note: F = Frequency/number of farmers; – = this reason not given.

Table 19.3 Farmers' perceptions of advantages of major traditional soil management practices

| | Traditional practices | | | | | | | | | | | |
| Advantages | Mulching | | Household waste | | Crop rotation | | Intercropping | | Rough tillage | | Fine tilth | |
	F	%	F	%	F	%	F	%	F	%	F	%
Practice not labour intensive	17	15.5	53	48.6	49	53.3	24	26.7	7	6.7	14	14.6
Practice not expensive	19	17.3	43	39.4	28	30.4	47	52.0	33	30.0	17	17.7
Soil fertility improvement	40	36.4	60	55.0	42	45.7	–	–	13	11.8	–	–
Weed control	62	56.4	20	18.3	–	–	–	–	–	–	12	12.5
Conserve water	25	22.7	–	–	–	–	–	–	–	–	–	–
Increased crop yields	30	27.3	42	38.5	45	48.9	–	–	15	13.6	–	–
Aids good germination	–	–	–	–	–	–	–	–	–	–	40	41.7
No inputs required	–	–	40	36.7	–	–	–	–	21	19.1	32	33.3
Local materials used	32	29.1	45	41.3	–	–	–	–	–	–	–	–

Note: F = Frequency/number of farmers; – = this reason not given.

Table 19.4 Farmers' perceptions of disadvantages of major traditional soil management practice used

| | | | Traditional practices | | | | | | | | | |
| Disadvantages | Mulching | | Household waste | | Crop rotation | | Intercropping | | Rough tillage | | Fine tilth | |
	F	%	F	%	F	%	F	%	F	%	F	%
Expensive	22	20.0	–	–	–	–	–	–	23	20.9	2	2.1
Labour intensive	43	39.1	15	13.8	–	–	–	–	11	10.0	29	30.2
Mulch destroyed by pests	6	5.5	–	–	–	–	–	–	–	–	–	–
Limits water infiltration	2	1.8	–	–	–	–	–	–	–	–	–	–
Leads to pests and diseases	–	–	44	40.4	–	–	–	–	–	–	–	–
Inadequate knowledge	–	–	–	–	13	15.2	–	–	–	–	–	–
Low yields	–	–	–	–	–	–	41	45.6	–	–	–	–
Time consuming	–	–	–	–	–	–	–	–	17	15.5	13	13.5

Note: F = Frequency/number of farmers; – = this reason not given.

Farmers' perception of the advantages of mulching also related to affordability of the practice in terms of availability of local material used, with 29.1 per cent of farmers using mainly mulch from bananas and trash from cultivated fields (Table 19.3). However, a somewhat contrary finding was that the majority of farmers who bought mulching materials reported disadvantages of mulching as being labour intensive (39.1 per cent) and expensive (20 per cent) (Table 19.4). Given that they were likely to carry mulching materials from long distances away and the fact that most banana plantations are near homesteads, other alternative sources of mulch, especially grass cut from swamps, meant investing in additional labour.

Household waste utilization

Household waste utilization involves disposing of household waste – especially rubbish accumulated during cleaning and food remains – by scattering it in gardens (vegetable gardens and banana plantations), usually close to homesteads. The majority of farmers used this to improve soil fertility (57.8 per cent) and control weeds (24.8 per cent). Other perceived advantages of using household waste in this way included no requirement for inputs (36.7 per cent), use of local materials (41.3 per cent), low labour requirements (48.6 per cent), and lack of expense (39.4 per cent). In this respect, farmers perceived use of household waste as affordable and undemanding. In terms of disadvantages, 40.4 per cent of farmers reported that household waste had a negative effect on banana plantations because it led to pests and disease (banana weevils and panama wilt). However, this concern was mainly due to some farmers' lack of knowledge of the correct handling and application of waste, which greatly reduces the problems.

Crop rotation

Farmers used crop rotation because of its positive effect on soil fertility (46.7 per cent) that consequently led to increased crop yields (53.3 per cent). This practice was also considered non-labour intensive (53.3 per cent) and inexpensive (30.4 per cent). This implied that the practice was perceived to improve crop production with only minimal requirements within the means of most farmers. The only disadvantage reported was inadequate knowledge of the practice among farmers. Since crop rotation was considered a traditional practice, individual farmers used various crop sequences depending on their objectives, regardless of the proper sequences that are beneficial.

Intercropping

Although intercropping was recognized as a major traditional soil management practice, the majority of farmers used it for reasons other than soil improvement. These were increased food security (84.4 per cent) and inadequate land (24.4 per cent). Farmers perceived intercropping as a means of meeting the primary objective of improving crop yields and obtaining food for household consumption as well as a surplus for income to meet other family demands. The fact that this area has a rapidly growing population explains the problem of inadequate land. As a result, reduced land sizes are susceptible to soil erosion, leading to declining soil fertility which in turn leads to low crop yields. This poses risks of crop failures and so farmers opt for intercropping in order to spread risks by cultivating several different crops. This implies that farmers perceived the contribution of intercropping to increased household food security as key to subsistence-level production. At the same time, farmers perceived intercropping as inexpensive (52 per cent). This may be accounted for by the fact that growing more crops on a single plot is less demanding of time and labour, certainly far less than growing those same crops on separate plots. But some farmers (45.6 per cent) reported one of the disadvantages of intercropping as being low crop yields. This discrepancy may be explained by the fact that some farmers use intercropping without considering other agronomic practices such as timing, spacing, and appropriate crop combinations. The positive aspects of intercropping are only achieved with attention to these additional crop management practices.

Rough and fine tillage

The majority of farmers used the traditional methods of rough tillage (42.7 per cent) and fine tilth (44.8 per cent), which are characteristic of annual cropping systems. These practices were also used because they were perceived to increase yields. Unlike fine tillage, rough tillage was perceived to lead to improved soil fertility (11.8 per cent) because it involves mixing grass and trash with soil during cultivation. Usually the field is left for some months after rough tillage before the second ploughing, while the grass and trash in the process decay, hence increasing organic matter in the soil. As a result crops grown on these fields do well. Also, this practice was found to be inexpensive, as it did not require inputs (19.1 per cent). Fine tillage was reported to require no inputs and to enhance good germination (41.7 per cent) and assist in weed control (12.5 per cent). Given that fine tilth is also useful for planting small-seeded crops such as millet and sorghum, important annual crops for the area, a fine seedbed is indicated for the germination. Perception of fine tillage as a method of controlling weeds is related to reducing weed seeds because trash is buried, removed, or burnt, leading to fewer weeds. The proportion of farmers who perceived fine tillage to be labour intensive (30.2 per cent) was bigger than those who perceived it not to be (14.6 per cent). This may be attributed the status of the field being tilled. A virgin field or field under fallow requires more labour to prepare a fine-tilled seedbed. If the field has been previously cropped, less labour is needed. Rough tillage includes the expense of hired labour. As most households use family labour, 30 per cent of farmers perceived the practice to be inexpensive, although time-consuming. Unexpectedly, farmers who perceived rough tillage to be time-consuming (15.5 per cent) were more than those for fine tilth (13.5 per cent).

Reasons for using major introduced soil management practices

Table 19.5 presents the reasons given for using the major introduced soil management practices, while Tables 19.6 and 19.7 summarize the perceived advantages and disadvantages.

Trench and soak pit utilization

The majority of farmers used trenches because they were effective in both water conservation (71.9 per cent) and erosion control (63.2 per cent). However, even though trenches were perceived to conserve soil and water, the practice has negative aspects such as labour requirements (61.4 per cent) and expense (35.1 per cent). The typical topography of the area explains farmers' reasons for using these practices. Because of

Table 19.5 Farmers' reasons for using the major introduced soil management practices

| | Introduced practices | | | | | |
| Reasons for utilization | Trenches | | Soak pits | | Controlled bush or trash burning | |
	F	%	F	%	F	%
Erosion control	36	63.2	–	–	–	–
Conserve water	41	71.9	16	59.3	–	–
Bush clearing	–	–	–	–	12	28.6

Note: F = Frequency/number of farmers; – = this reason not given.

Table 19.6 Advantages of major introduced soil management practices

| | Introduced practices | | | | | |
| | Trenches | | Soak pits | | Controlled bush or trash burning | |
Advantages	F	%	F	%	F	%
Practice not labour intensive	–	–	–	–	14	33.3
Practice not expensive	–	–	5	18.5	25	59.5
Conserve soil and water	38	66.7	15	55.6	–	–
Increased crop yields	–	–	–	–	5	11.9
No purchased inputs required	–	–	–	–	28	66.7

Note: F = Frequency/number of farmers; – = this reason not given.

Table 19.7 Disadvantages of major introduced soil management practices

| | Introduced practices | | | | | |
| | Trenches | | Soak pits | | Controlled bush or trash burning | |
Disadvantages	F	%	F	%	F	%
Expensive	20	35.1	2	7.4	–	–
Labour intensive	35	61.4	23	85.2	–	–
Difficult to make	2	3.5	–	–	–	–
Land dries quickly	–	–	–	–	4	9.5
Reduced soil fertility	–	–	–	–	8	19.0

Note: F = Frequency/number of farmers; – = this reason not given.

the hilly terrain with cultivation and settlement over the entire landscape, especially in Mwizi subcounty, use of trenches become inevitable. In addition, trenches were utilized in banana plantations for both food and income purposes related to maintenance of soil fertility. Retaining fertility and water by trenching meets farmers' primary objective of increasing yields, even at considerable cost.

Soak pits had the same functions as trenches, especially in conserving water (59.3 per cent). Soak pits were inexpensive (18.5 per cent) and conserved soil and water (55.6 per cent), according to farmers. Some farmers, however, who used hired labour perceived soak pits as expensive (7.4 per cent). The majority perceived the practice as labour intensive (85.2 per cent), possibly because making the basin-like pits involved scooping considerable amounts of soil from the pit.

Controlled bush/trash burning

Although controlled burning was considered an "introduced" soil management practice, farmers used it for non-soil-related reasons, mainly because it was a traditional method of bush clearing (28.6 per cent). Land clearing involves burning limited patches of land or trash left over from cultivation. This was perceived as good for yields and easy to operate, requiring no purchased inputs (66.7 per cent), little expense (59.5 per cent), and low labour requirements (33.3 per cent). However, some farmers found the practice negative to the soil, reducing fertility (19 per cent) and causing drought (9.5 per cent). The negative effect on soil fertility is accounted for by destruction of organic matter and soil structure. However, farmers who thought the practice led to increased crop yields were likely to have used the practice on millet. The growth of millet flourishes with released nutrients, especially potassium, as a result of burning. The perceived value may be due to overall vegetative growth in addition to actual grain yield. Farmers tend to observe and evaluate the practicability of practices they use.

Conclusion

The appraisal of different soil management practices has revealed that farmers have complex and dynamic reasons for utilizing techniques based on the effectiveness of the practice in solving agricultural problems and increasing crop production at low cost. Understanding farmers' criteria when evaluating different soil management practices assists in promoting relevant options for the sustainable enhancement of soil quality and crop production.

REFERENCES

Brookfield, H. 1989. "The human context of sustainable smallholder development in the Pacific Islands", in Proceedings of the International Board for Soil Research and Management (IBSRAM), No. 8, *Soil Management and Smallholder Development in the Pacific Islands*. Bangkok: IBSRAM.

Busingye, P., J. K. Tumuhairwe, and E. N. Nsubuga. 1999. "Factors influencing farmers' decisions to use mulch and trenches in the banana plantations in Mwizi, Mbarara district", in *Proceedings of the 17th Conference of the Soil Science Society of East Africa (SSSEA)*. Kampala: SSSEA.

ICRAF. 1997. *Maintenance and Improvement of Soil Productivity in the Highlands of Ethiopia, Kenya, Madagascar and Uganda*. African Highland Initiative

Technical Report Series No. 6. Nairobi: International Centre for Research in Agroforestry.

Loevinsohn, M. and F. Wangati. 1993. *Integrating Natural Resource Management Research for the Highlands of East and Central Africa.* Nairobi: International Centre for Research in Agroforestry.

Opio-Odongo, J. 1999. "Applied soil science research in a changing society: Challenges for Uganda soil scientists in the new millennium", in *Proceedings of the 17th Conference of the Soil Science Society of East Africa (SSSEA).* Kampala: SSSEA.

Tumuhairwe, J. K., C. Nkwiine, E. N. Nsubuga, and F. Kahembwe. 1999. *Agrodiversity of the Banyankore and Bakiga People of South-western Uganda.* Tokyo: United Nations University Press.

Part IV

Policy recommendations

20

Technical and policy recommendations for sustainable management of agricultural biodiversity: Recommendations for Tanzania with contributions from meeting participants

Fidelis B. S. Kaihura and Deusdedit M. Rugangira

Introduction

Smallholder farmers throughout the tropics are adept at using the natural diversity of the environment for managing the soil, water, land, and vegetation, and for production. Arguably, they have conserved more biological diversity and more economically important species than all protected areas combined. This is undoubtedly the case in Tanzania, where the highest levels of natural biodiversity coincide with some of the highest population densities, as on Mount Kilimanjaro and the adjacent PLEC sites on Mount Meru. Smallholder farmers have systems of land use and practices that have stood the test of population growth and environmental change. Among smallholder farmers there is a large untapped source of knowledge which could potentially contribute to key policy issues on the international agenda, such as:

- conservation of biological diversity, especially of agricultural biodiversity in food, medicinal, and livelihood-support plants
- protection of globally important land-use systems, the so-called "ingenious heritage" systems that display land management practices developed over centuries of farmer experimentation and handed-down knowledge
- control of land degradation through the enhanced protection afforded to the soil by biodiverse plant associations
- food security and sustainable rural livelihoods.

PLEC calls this agrodiversity (see Chapter 2 for an illustration of the PLEC agrodiversity framework). It promotes agrodiversity as a means of supporting global objectives of biodiversity conservation while also supporting human needs and development. Through the project activities, over 200 scientists and many times more collaborating farmers are discovering just how and why agrodiversity is important for future sustainable development.

Generating recommendations

During the period March 1998 to January 2002, annual project workshops and six-monthly feedback meetings were conducted with PLEC farmers and other stakeholders in Arumeru. After every workshop or feedback meeting a number of specific observations and recommendations were developed. Some of those that needed implementation at village and district levels are currently being put into practice. Those that need to be incorporated into national plans and/or budgets were left to policy-makers for consideration.

Technical recommendations

Climatic change

Rainfall data analysis for the period 1953–1997 for the Arumeru district indicated a general decrease in total rainfall and length of growing seasons as well as changes in the onset and cessation of rains. Farmers also supported the findings that occurrences of drought and length of dry seasons had increased. This evidence of climatic change suggests a number of policy issues.

Creating awareness

A national strategy should be developed to create awareness among farmers and land users of the trend revealed in such studies. Alternative strategies to cope with the situation should also be outlined. Awareness of deep ploughing for soil moisture conservation and availability of appropriate implements should also be created. Deliberate efforts should be taken to document what farmers are doing in order to work towards their improvement and wider application. For example, water-harvesting techniques and deep-ploughing experiments with and without manure were found to increase maize yield by over 30 per cent. Such techniques were recommended for adoption in semi-arid environments. Current coping strategies used by farmers include the use of locally developed drought-, pest-, and disease-tolerant seeds with many other multi-purpose uses, diverse cropping systems, area/farmer-specific crop rotations, and traditional irrigation systems. Awareness creation would not only alert policy-makers to the viability and productivity of farmers' strategies, but also help to disseminate these examples of good practice to other farmers.

Support of farmer initiatives

The prevailing situation in Arumeru prior to PLEC was that outside commercial interests would exploit the forests, leaving no benefit to the villages and local people. Farmers' current initiatives in biodiversity conservation and rehabilitation of degraded lands should be recognized, supported, and rewarded at all levels. National institutions that are immediate beneficiaries of farmer conservation initiatives include:

- the Tanzania Electrical Supplies Company (TANESCO), through conservation of catchments and more reliable river flow and hydroelectrical power generation
- the National Urban Water Authority (NUWA), through cleaner and more reliable water supplies from mountain catchments
- wood and timber industries, through access to sustainable sources of supply.

Such initiatives should return some of these benefits to rural communities as incentives and use for rural community development. Specific reference was made to conservation of water catchments, natural forests, wetlands, lakes, hot springs, and game reserves. Similarly, natural forests conserved by the government should benefit the villages surrounding them. In policy terms, support for farmers rather than outside commercial interests, which return nothing to the local community, would bring substantial national benefits in protecting ecosystem service functions as well as in local rural development.

Unsustainability of current practices

Current practices of clearing natural forests and planting commercial forests should be revealed for their unsustainability and inability to support the national and local economies. Afforestation should be directed to areas of little vegetation. Deliberate large-scale afforestation of semi-arid lowland environments with particular emphasis on indigenous trees should also be initiated by responsible institutions. Meanwhile, the potential of indigenous tree species and pastures, particularly with respect to biodiversity conservation, fertility restoration, erosion control, and their use in agroforestry systems, should be emphasized. Most afforestation in Arumeru is directed to humid and sub-humid areas and concentrates on clearing natural forests in order to replace them with planted exotics. This reduces biodiversity and agrodiversity benefits in smallholder farms. Established plantations should avoid monocultures and include indigenous trees.

Encouraging conservation at household level

Conservation of established woodlots at household level should be encouraged and strengthened. Both indigenous and exotic tree species should be planted. Woodlots and the utilization of multi-purpose woody species have been shown to be more effective in income generation and agrodiversity conservation (especially in semi-arid environments) than crop production. They also provide food to both people and livestock.

Unreliable weather forecasting

One issue highlighted by land users is the poor state of weather forecasting from national-level institutions. Good rainy seasons have sometimes been forecast, leading to farmers making appropriate cropping decisions, only for them to be disappointed by unexpected droughts during the rainy season. Similarly, poor seasons have been forecast and farmers have taken action to plant drought-tolerant species at wide spacing, only to have crop failures in the unpredicted heavy rains. As recording stations are scarce and some are non-functional, it is recommended that more modern and efficient instrumentation should be established for key sites such as local schools, churches, and hospitals in addition to current sites. Recorders need to be trained to take timely and accurate measurements. An adequate network of meteorological stations with functioning equipment and trained staff is essential as a nationally provided service. It would enable farmers to make informed decisions in their management of agricultural biodiversity, with lessened risk of failure and inability to protect rare varieties. However, it must be noted that even where weather forecasting is good and operates with considerable funding and technical assistance, it is still only an exercise in probability, which many farmers (especially in developing countries) can find a difficult notion when trying to make agricultural decisions.

Improved dissemination of weather information

Further to the issue of the reliability of weather forecasting, there is also a question of access to this knowledge. Current weather forecasting by radio does not reach the majority of small-scale farmers because radio receivers are few and signal strength insufficient to reach remoter rural parts. Alternative media should be investigated, such as pamphlets, church gatherings, public meetings, and publicity in schools and hospital areas, in order to reach the majority of the farmers in need of weather information. The meteorological department should also send information to other ministries in order to assist in spreading the messages.

Use of drought-tolerant crops

Eating habits and food preferences sometimes run counter to the interests of biodiversity, where specific varieties may hold dominance in local people's choice of food. This is particularly the case for drought-tolerant crops such as sorghum, millet, cassava, and sweet potatoes. Some of these crops and local varieties used to be commonly grown, but have largely now disappeared. Researchers should interact with farmers to document desired qualities for improved seed in their breeding programmes (e.g. taste and colour of *loshoro* for *waarusha* and *wameru*). Drought-tolerant improved seeds and local potential seeds and appropriate cropping systems were recommended for the semi-arid zones.

Soil fertility improvement and biodiversity

Soil fertility in densely populated areas of Arumeru depends on nutrient imports from support plots (i.e. plots in other parts of the landscape, sometimes more degraded than the plots to which the nutrients are applied) through the transfer of crops and crop residues. However, support plots are seldom well managed. Some are in semi-arid erosion-prone areas and face losses of biodiversity and rapid land degradation. The following recommendations were made.

Conservation farming

Conservation farming methods and the use of fertilizer inputs should be promoted to improve soil fertility and prevent loss of biodiversity. Leguminous plants may be used to strengthen physical conservation structures of support plots, especially on steep slopes. Both conservation farming and integrated soil fertility management programmes should be launched for sustainable production of support plots in Arumeru and densely populated areas elsewhere. The programme should include effective use of crop residues, use of leguminous cover crops such as *Mucuna* spp. and *ng'wara*, nitrogen fixation through cereal/legume intercropping or mixed cropping, agroforestry techniques, and proper uprooting of legume stovers from farms. Indigenous trees and pastures should be used in fertility restoration and rehabilitation of degraded lands as a fertility restoration and indigenous agrodiversity conservation strategy for both the main production plot and the support areas away from the homestead.

Incentives

The role of incentives, whether as positive encouragement for practices or penalties applied through by-law provisions, should be considered carefully. For example, rewards for planting trees of biophysical significance and biodiversity enhancement could be developed as an incentive for effective use and management of natural resources. Rewards and incentives through local laws could also apply to discouraging practices such as the plugging of gullies with valuable manure, the piling of ashes after burning, or placing manure along roadsides. Through district sustainable resource management programmes implemented through existing NGOs and extension staff, much could be achieved to enhance biodiversity, protect the environment, and promote rural livelihoods.

Studies of traditional methods

Indigenous technologies need to be examined more closely for their scientific rationale and their applicability to other communities and situations. Technologies and land-use strategies found in PLEC-Tanzania include the shifting of family houses to different parts of the landscape, cattle pens for stall-feeding, and the location of toilets. Crop rotations and cropping systems for fertility improvement also demonstrate positive aspects in the sustainable use of the natural environment. Promising elements can then be improved, if appropriate, to be more effective based upon a scientific understanding of the complex interactions.

Increase farm incomes

Use of fertilizers is essential if improved seed is used and proper crop husbandry practices of pest and weed control are employed. However, farmers are adversely affected by structural adjustment policies that saw the removal of subsidies on inputs, especially for small-scale farmers who have little purchasing power. The government should initiate and finance rural development projects that aim to increase small-scale farmers' income as well as improving biodiversity.

Indigenous technologies – their development and transfer

The PLEC activities recognize farmers' indigenous knowledge and their associated diverse and dynamic coping strategies with regard to resource management and production. In addition PLEC recognizes that rural livelihoods and food security have greatly depended on the experience and knowledge of these farmers. Unfortunately, the most successful indigenous resource management skills are only known by the older generations, and none of this knowledge is documented for future custodians of the land. In addition, female farmers (who are primarily responsible for land management) have received little attention from extension activities or training. The following recommendations are therefore made.

Farmers training farmers

Expert or skilled farmers should be empowered and facilitated to undertake farmer-training programmes. There should be a focus on successful resource management relevant to the area's farming/cropping systems and biodiversity conservation. Researchers should also be trained to understand expert farmers' production systems and integrate their current methods with scientific methods for their improvement. In this method both researchers and extension workers facilitate the process.

Transfer of knowledge between generations

A system of transfer of indigenous knowledge from older to younger generations should be established and reinforced. Knowledge of the names of different types of soils, local-language soil quality indicators, and uses of different shrubs, trees, and grasses are some examples of information that should be shared. The process may include documentation and incorporation of indigenous knowledge of resource management in primary and secondary schools. The government may also consider financing income-generating and biodiversity-enhancing activities in the rural communities that will occupy young farmers and prevent them from moving to towns. Such activities have been initiated by PLEC in Arumeru. Similarly, temporal and spatial changes in resource management and their effects on agrodiversity and environment should be taught in schools at all levels.

Gender awareness

The PLEC project has shown how vital the knowledge of both women and men is in the conservation and sustainable use of biological diversity. An element of gender balance should be incorporated in community development and resource management programmes at all levels, from household to national. Support of women's groups for agrodiversity enhancement and conservation should be initiated through existing NGO rural development programmes. To be successful, women who are expert in resource management should pilot the process. Self-managing and self-financing groups should be formed, especially for women. Each group should address a single environmental management activity that contributes to agrodiversity and increases income.

Exhibitions for expert farmers

Expert farmers should have their successes recognized and be given the chance to show the agricultural products arising from their own initiatives. Their exhibitions should cover agricultural, livestock, and forestry products and successful resource management models. The current practice is that farmers are invited to see what others have done and adopt the practices if they can.

Markets for diverse crops

The government should seek markets for the diverse crops produced to act as an incentive to farmers' agrodiversity management. The pricing mechanism of farm products should be harmonized to encourage farmer initiatives.

Linkage

It was observed that the current research extension linkage is weak and that outreach programmes, seminars, and workshops were better tools in technology development and dissemination. They are, however, expensive tools. Most research outputs remain as office documents and never reach farmers, despite the long-running farming systems research (FSR) approach. Farmers – even those in the neighbourhoods closest to research centres – do not know what research is being carried out. As a result, farmers may carry out parallel research on their own farms to address their production problems. Such work (most of which is useful) is neither documented nor shared with neighbouring farmers or researchers. Stakeholders also noted that most extension work is unplanned. The following recommendations were made.

Outreach programmes

The government should fund outreach programmes for farmers, researchers, and extension staff to enhance continuous interactions of experts for focused and cost-effective research and technology development for both on-station and on-farm research.

Leading role of expert farmers

In rural technology development and dissemination, expert farmers should be empowered to lead the process, facilitated by researchers and extension staff.

Support of local leadership

District and village leadership should give support and necessary information to extension staff. Leaders should also avoid assigning them non-professional engagements that undermine their linkage and extension activities, such as levy and tax collection.

Collaboration between farmers and researchers

Collaboration and interaction between farmers and extension staff and researchers should be maintained, as is the case in Arumeru under PLEC. However, in order to do so, all parties should be facilitated and motivated. Similarly, farmers, relevant stakeholders, and extension staff should develop their work-plan together, at least on a quarterly basis. All involved parties should respect the resulting work-plan and appointed authorities should monitor achievements and address setbacks during implementation.

Migration

It was observed that in densely populated areas such as Arumeru, farmers are advised to move to identified free land – for example, Kiteto. However, there are usually no prior arrangements for land-use planning of the identified new land. Some farmers have to return, or else live with many difficulties. The following recommendation was therefore made.

Evaluate and plan land use

For all areas with potential for accommodating incoming migrants, there is a need to carry out an evaluation of the land and establish a land-use plan for the expected newcomers. Farmers need to be trained in the potential and constraints before working on the new land, including the advantages of protecting the natural biodiversity and conserving the agrodiversity.

Access to agricultural inputs and advisory services

Despite the existence of agricultural input services, farmers face problems of reaching town in time and identifying shops that render suitable services (e.g. the inputs they require such as seeds, fertilizer, pesticides, and insecticides). They also lack proper guidance on the application of chemicals in a safe and timely manner. This includes application of manure, a readily available input in semi-arid environments. They also do not have any indications of prices. Sales staff often give incorrect information on the use of agricultural inputs to farmers, and in some cases input recommendations such as spacing do not conform with local cropping systems. The following recommendations were therefore made.

Bring services closer to farmers

The government should bring agricultural input services closer to farmers, preferably to village centres. NGOs and input stockists could be contracted to provide soft loans on transportation of agricultural inputs to villages. The package size of provided inputs should bear farmers' purchasing power in mind and prices should be known. Sales people should have an agricultural training in order to advise farmers appropriately. Similarly, dealers in agricultural inputs should prepare brochures giving application details of the most common inputs. The government should increase the budget for the Ministry of Agriculture and Food Security to embark on training farmers in input use. A common complaint among farmers was that advice was inadequate and omitted some basic information; for example, some farmers cannot differentiate between triple-superphosphate and calcium ammonium nitrate, yet their content and applicability are very different.

Create awareness

As covered earlier with regard to the onset of rains and timely planting, awareness creation should also be extended to include information on timely application and incorporation of inputs and their best management for increased yields, particularly in semi-arid environments.

Understanding local cropping systems

Distributors of maize seed should bear in mind the cropping systems of regions they serve. In Arumeru, for example, maize is always intercropped with beans. Recommendations should therefore take spacing into account. Current recommendations are for sole maize, and this has forced farmers in Arumeru to embark on their own research to establish a proper intercropping spacing.

Policy recommendations

Implementation of several CBD articles

Tanzania ratified the Convention on Biological Diversity (CBD) in 1994. The CBD promotes conservation of biological diversity, equitable sharing of benefits derived from the use of genetic resources, and the sustainable use of its components. For the convention to be operational in each country, policies and legislation for its implementation need to be put in place. However, Tanzania has not finalized policy and legislation formulation regarding implementation of the CBD and the related articles such as the intellectual property rights (IPR), access and benefit sharing (ABS), and genetic resources (GR) provisions. It was also noted that current national policies (e.g. the National Environmental Policy of

1997, the National Land Policy of 1995, and the National Agricultural and Livestock Policy of 1997) do not mention agrodiversity or agricultural biodiversity.

There have been several government initiatives regarding environmental management and biodiversity conservation. These include:

- establishment of the National Plant Genetic Resources Committee
- the 1990 Plant Genetic Resources and Biotechnology workshop recommendations
- the 1994 National Environmental Action Plan
- the National Environment Management Act No. 9 which established the National Environmental Management Council
- establishment of the National Commission for Science and Technology (especially its National Research Clearance Committee).

However, the legislative process to introduce several of the CBD articles has not yet been completed. Farmers expressed concern that they are among the most disadvantaged, as their indigenous knowledge and practices and indeed indigenous genetic resources are being exploited without legal controls. Technologies developed by local farmers to use indigenous tree species to control pests and diseases in crops and livestock are a case in point. Outsiders are seeking details of species and practices without promising any benefits in return to farmers. There are no provisions for access or transfer of knowledge to outside parties with commensurate recompense to the local custodians of biodiversity.

Members and stakeholders of PLEC and its activities recognize that agrodiversity covers more than just monitoring and conservation of biodiversity. It takes into account the people who manage the diversity, the management systems, the biophysical environment on which plants grow, and the organizational aspects influencing conservation and management. Indeed, it supports human needs and development. The recommendations are as follows.

Improve local knowledge of the CBD

The government should be advised on the need to create/strengthen village committees on the environment and enhance awareness of the CBD. This will facilitate policy implementation in aspects of benefit sharing in all the provisions of the convention, as noted above. End users of environmental conservation products, such as commercial interests in water, timber, and wildlife, should be able to pay back part of the earnings generated to the farmers as incentives.

Creation of a national database

A national database on biodiversity should be established and shared between relevant ministries. The format and methodology for data collection should be harmonized and distributed to institutions in need. All national biodiversity projects should contribute to this database.

By-laws to protect small-scale farmers

At the community level, village governments need to establish by-laws to protect small-scale farmers' indigenous knowledge and practices. They should also establish legislation to support community-based decisions and by-laws in agrodiversity conservation, particularly those associated with access and benefit sharing. Established village by-laws should be recognized by the central and local governments to assist enforcement.

Farmer input to plant breeding

Special attention should be paid to understanding and documenting farmer-desired qualities when breeding improved seed and other germplasm. Botanical gardens should be revived at research and training institutes where traditional crops, livestock, indigenous agroforestry trees, and pastures are maintained and eventually reintroduced into modern land management and cropping systems to enhance lost agrodiversity. The process should include recording and monitoring individual farmers' own botanical gardens, some of which in Arumeru are truly remarkable.

Knowledge exchange between farmers

The government should be advised on the need to support and facilitate the exchange of knowledge of agrodiversity between farmers. Through a well-managed database of the various local seeds kept by farmers, they would be able to exchange information between areas. Current importation or transfer procedures constrain on-farm agrodiversity enhancement and conservation. Researchers should also recognize and utilize farmers' indigenous knowledge.

Strengthen cross-sectoral groups

For purposes of mainstreaming, the government should strengthen cross-sectoral multidisciplinary agrobiodiversity coordination groups. Such groups should be established or existing ones expanded to include agro-diversity professionals and administrators in order to speed up the legislation process and monitor implementation of established legislation.

Implementation of the AU Model Law

The Organization of African Unity's (now named the African Union, AU) Model Law of 2000 was developed for the protection of the rights of local communities, farmers, and breeders. It was also designed to regulate access to biological resources in order to protect Africa's common heritage of biological diversity and the livelihood systems dependent on it. It provides the necessary framework for member states of the AU to draft specific national legislation consistent with their political orientation, national objectives, and level of socio-economic development. This is because the growing forces of global trade are seeking to secure a monopoly over Africa's valuable biodiversity, knowledge, and markets through the guise of global and bilateral trade agreements, which are intrinsically unfair. The AU Council of Ministers specifically recommended that African countries develop national laws, as well as regional regimes and common negotiating positions in international law and related issues, to protect Africa from this onslaught. This legislation should enshrine the right to continue living according to ecologically coherent practices and to establish a boundary beyond which monopolies cannot penetrate. The following quote, taken from the January 1999 African regional workshop on understanding biodiversity-related international instruments regarding benefit sharing and genetic resources, illustrates the need for such legislation.

The Convention on Biological Diversity assumes that when a state allows access to a sample of genetic resources, it is in return entitled to insist on a number of benefits. Research activities on the genetic resources that country provides have to be done in its territory to help that country build capacity. All the information generated by research on that genetic resource is subject to repatriation. Any biotechnology applied on the genetic resource must be accessible to the country that provided the genetic material. A fair and equitable share of benefits accruing from the use, including from commercial gains of the genetic resource must also be given to that country. But all this is conditional upon a mutually agreed contract. To our understanding, there is as yet no African country with the appropriate legislation to cover such contracts.

The industrialised countries know this, and many of them have been undertaking expeditions to Africa to collect genetic resources before African countries wake up to enforce their sovereign rights over these resources. As usual we wake up after the thief has taken what he wants and has gone away.

The following recommendations were made.

Use of the AU framework

The AU model framework should be used to develop appropriate legislation in favour of the custodians and managers of biodiversity in Tanzania – rural communities, farmers, and breeders.

Establishment of a cross-sectoral team

A cross-sectoral team should be established to develop effective *sui generis* systems or a combination to protect plant varieties and livestock breeds from open exploitation. Specifically, the problems associated with lack of legislation in the informal seed sector (e.g. farmers' varieties and breeds) versus the commercial sector (e.g. improved varieties/breeds) should be addressed. Zimbabwe, for instance, has developed country standards to qualify local varieties. In this case, local seeds widely used by farmers and with locally desirable potentials that meet locally set standards are termed "quality declared seed" versus "certified seed". Similarly, farmers' own local seed – selected and tested over time and used widely – should be recognized officially in accordance with national or zonal standards for "quality declared seed", multiplied, and studied further in order to become sources of desirable traits in breeding for certified seeds.

Support of poor farmers' conservation of biodiversity

PLEC-Tanzania activities have revealed that the highest biological diversity was on poor farmers' fields rather than on those of rich farmers. Also, fragile semi-arid environments have a richer biodiversity than less fragile humid and sub-humid environments. Poor farmers are also primarily responsible for the regeneration of endangered species. However, less attention has been paid by agencies to both fragile environments and poor farmers, particularly in semi-arid environments, in support of their endeavours in environmental and biodiversity conservation. The following recommendations were made.

Government to recognize the role of smallholder farmers

The government should be advised to recognize and support poor smallholder farmers' efforts at conserving biodiversity and regenerating degraded environments in semi-arid environments. PLEC researchers observed that most funding has been directed to areas with the most agricultural potential and where research and demonstration will obtain quick results. There is a need to pay more attention to farmers in semi-arid environments.

Conservation of traditional ecosystems

The majority of farmers in Tanzania depend on traditional varieties and ecosystems rather than hybrids and commercial crops. While working on improved methods to increase productivity and production, traditional ecosystems' management should be well conserved as a major source of inputs in producing improved breeds and developing improved management methods.

Documentation of indigenous species

Because of a lack of documentation and in-depth studies of the potentials of indigenous trees, fodder, pastures, and local animal breeds, current programmes on environmental conservation and rehabilitation of agricultural lands emphasize the use of exotic trees, fodder, and pastures. They ignore, or are not aware of, the potential of indigenous species with regard to fertility restoration, land rehabilitation, and application in agroforestry systems. Documentation and studies of indigenous species and their role in environmental conservation should therefore be initiated. Farmers should also be encouraged to use indigenous trees in combination with introduced trees in forests, farms, and pastures on both individual and government lands and farms.

Conservation by-laws

By-laws should be developed by rural communities on the conservation of biodiversity through appropriate management of indigenous and planted woodlots. Woodlots are a potential source of income, especially in semi-arid environments with unreliable rainfall.

Conclusion

PLEC-Tanzania has worked in Arumeru district for some seven years and has investigated the whole subject of agrodiversity, but from a farmer perspective. The demonstration sites have been used to bring farmers, researchers, and policy-makers together. In its findings, PLEC has revealed a number of key areas where policy could and should address areas of concern to both local people and national agencies, in addition to global policy-makers concerned about environmental change. These key areas, listed in the body of this chapter, reflect service provision (e.g. weather forecasting), education and training, and awareness creation. In the final analysis, the main policy issue is how to raise the profile of the importance of agrodiversity and farmers' management of it in national and local planning. PLEC-Tanzania has demonstrated the importance; it is now up to politicians and local people to pursue these concerns and see legislation and policy instruments that support biodiversity.

21

Developing policy and technical recommendations for the conservation of agricultural biodiversity in Uganda

Joy K. Tumuhairwe, Charles Nkwiine, John B. Kawongolo, and Francis K. Tumuhairwe

Introduction

Uganda was one of the first countries to sign (1992) and ratify (1993) the UN Convention on Biological Diversity (CBD), because the Ugandan people and government acknowledge the many values and benefits from biodiversity. The ratification indicated Uganda's commitment to international policy, and posed the challenge to both policy-makers and professionals of how to implement the CBD provisions into national and local policy. Unfortunately, in so far as agrodiversity is concerned, the majority of efforts for biodiversity conservation have been concentrated in protected areas, mainly national parks and forest reserves, which occupy only a very small proportion of the total land area and which contain probably only a small proportion of Uganda's biological diversity.

There is also a human developmental reason why the concentration on protected areas was unfortunate. Being an agriculturally based economy, protected areas cannot satisfy the rapidly growing and multiple demands of the population (including food, shelter, medicines, and clothing). Current agricultural policy and action plans do not explicitly emphasize the conservation of the diversity of biological resources, including management, which PLEC calls "agrodiversity" (see Chapter 2). Instead, the focus has been on increasing production through expanding acreage and selection of species and varieties, diversification of enterprises for broadening the economic base, and improving soil and crop management for high yields. The government also has a plan for the modernization of agriculture (PMA) as part of its strategy for the Poverty Eradication Action Plan (PEAP). The PMA emphasizes the use of improved, especially high-yielding, crop varieties and livestock breeds and market-oriented production, but does not highlight agrodiversity conservation. Conservation of genetic resources is promoted in *ex situ* programmes. However, there is a need to acknowledge the fact that sustainability of agrobiodiversity conservation is more certain when farmers are involved, and this is only possible *in situ* and where indigenous knowledge is tapped.

The Ugandan government has gone further to formulate a National Biodiversity Conservation Action Plan, but even the latest draft does not mention agrobiodiversity. The National Environment Management Authority (NEMA) is the implementing arm of government on all environment-related matters, including issues of biodiversity. However, even the regulations on managing hilly and mountainous areas stipulated in the Environment Statute are not implemented.

The NEMA is represented at district level, and it plans to set up local environment committees at grassroots level. It would therefore be in a position to work closely with agricultural and other relevant departments to assist rural communities in biodiversity conservation. Until now, there has been a distinct lack of technical and policy guidelines on conserving agricultural biodiversity.

The global benefits and even localized ecological values of conserving biodiversity are indirect and are therefore often not appreciated or valued by communities in the short or even medium term. However, it is timely now that efforts are made deliberately to conserve biodiversity on agricultural land. This would involve not only sustainable use, but also appreciation of the benefits of agrodiversity, protection of endangered species, and, where possible, cultivation of high-value species that would not otherwise survive on farmland.

One of the objectives of the UNU/PLEC project was to establish sustainable approaches to biodiversity conservation within the agricultural ecosystem. To measure the sustainability of an approach and establish how to ensure its sustainability is not easy in the short run. Practical indicators include conservation's acceptability to farmers and other stakeholders. Acceptability is measured by the willingness of farmers to experiment on the technology, to tell others about it, and, in the case of decision-makers and field practitioners, the willingness to incorporate recommendations from such experiments into their decision-making tools and implementation strategies. Involvement of all stakeholders, including farmers, in technology development, testing, and dissemination is PLEC's strategy to ensure sustainability of the developed technologies and approaches.

Among the many achievements of the PLEC project in Uganda is the development of a demonstration site in Bushwere parish, Mwizi sub-county, Mbarara district, with skilled expert demonstration farmers and motivated farmer groups and associations. During the process of developing the demonstrations, factors influencing conservation opportunities were noted and discussed with stakeholders. These were then used to develop some recommendations for use by policy-makers and field practitioners.

Devising policy and technical recommendations from the four years of PLEC is not only a way of ensuring sustainability of the developed approaches, but also a means of facilitating their wider adoption and utilization. Contributions to policy are the primary way of scaling up pilot project experiences to other communities.

The purpose of this chapter is therefore to share the experiences of the PLEC-Uganda team with both national and international stakeholders. The policy and technical recommendations for agrobiodiversity conservation in Uganda, developed through a series of processes involving stakeholders at the grassroots, local government, and national levels, are summarized. Although they are based on experiences from the Bushwere demonstration site, stakeholder involvement gave them wider applicability in Uganda and relevance to similar ecosystems elsewhere in the world. The goal is to develop policy strategies that integrate the objectives of agricultural production and environmental protection through land-use practices that conserve and sustainably manage agricultural biodiversity.

The objectives of making PLEC's technical work relevant to policy are as follows.

- To raise the main policy issues and suggest a national policy framework for sustainable use of agricultural biodiversity and its conservation.
- To provide technical guidelines for making agrobiodiversity conservation clearly beneficial and attractive to rural communities.
- To enhance accessibility of tools (legal and technical) for effective sensitization, decision-making, and policy implementation.
- To advocate the establishment of a national agrobiodiversity database and mechanisms for its regular monitoring and assessment.
- To provide technical guidelines for arresting and/or reducing erosion of agricultural biodiversity and other related land resources such as soil and water.

Summary of activities

Participatory technology development methods were employed throughout the four project years. The major steps towards constructing policy recommendations were as follows.

- Assessment of agrobiodiversity potential and conservation efforts.
- Evaluation of existing knowledge and technologies among farmers.
- Improvement of innovative farmers' practices.
- Demonstrations by innovative farmers to other farmers.
- Evaluation of developed technologies by relevant stakeholders through field workshops.
- Consultations on policy issues at community, district, and national levels – through personal contact and literature reviews.
- Findings were then subjected to a SWOT (strengths, weaknesses, opportunities, and threats) analysis during a technical workshop of the collaborating scientists through which important issues were raised and recorded by themes.
- Discussion of drafts with decision-makers, scientists, and farmers' representatives during a one-day technical workshop.
- Revision of the draft into acceptable (normal) policy document format, reducing the issues to only the most salient ones that can only be handled at policy level. During revision, the major issues were than sorted, limiting them to two or three per theme. The relevant recommendations were listed and only the feasible ones recorded. In some cases, explanations were added for clarity if necessary.

Summary of findings

- People deliberately conserve several species for direct food, socio-economic, and cultural benefits. Diversity is part of the normal strategy used by local people in constructing secure livelihoods.
- High agricultural biodiversity and related aspects of management still exist. Local people retain a high level of knowledge of agrodiversity, but may lack motivation and incentive to protect it.
- Policy guidelines on agrobiodiversity are lacking in the available government documents (even the recently drafted Biodiversity Conservation Action Plan).
- Local leaders lack access to legal and technical tools for effective sensitization, decision-making, and policy implementation.
- There is no database of agrobiodiversity in the country except that recently developed by PLEC, which specifically relates to demonstration site areas only. This situation hinders planning for monitoring and assessment of this valuable resource.
- Agrobiodiversity (being a multisectoral resource involving service functions supplied by government departments of agriculture, environment, forestry, economic planning, and industry) requires proper sensitization of the various interested parties for better coordination of people working in the relevant ministries.

Policy and technical recommendations

The resulting policy and technical recommendations are summarized under the following eight themes.

Theme one: Current status of agrobiodiversity

The important issues are detailed below.

- Most of the biodiversity on agricultural lands has restricted distribution to particular land uses. In the Bushwere demonstration site, for example, 60 per cent of plant species studied in 24 field types were recorded in only one or two of the seven land-use stages, and natural grasslands and bushlands occur only in small patches – contributing only 9 per cent each of total land cover. Natural grasslands have the fewest plant species, most of which are actually part of an anthropogenic fire climax association. Plants such as *Loudentia* spp. and *Cymbopogon* spp. rather than the natural biodiversity indicate erosion of biodiversity. Maintaining biodiversity requires careful conservation of species, habitats, and land uses, especially those species for which the community has direct use, e.g. *Xanthomonas sagitifolia* and *Solanum terminale*.
- Food crops with a high market value, such as Irish potatoes and garden peas, are often sold. This can mean that there is not enough for household consumption or seed provision for next season's crop. It also means undue attention and effort is paid to a restricted number of species to the detriment of agricultural biodiversity.
- Market forces of demand and supply, as well as changing tastes and preferences of younger farmers compared to those of older ones, have led to a drastic decrease in production of some important crops, for example millet and local varieties of Irish potatoes and beans.
- Population pressure, which increases demand for fuelwood and other land resources, has contributed to the disappearance of some highly beneficial trees such as *Combretum* spp. and *Ekikoyoyo* as well as some grasses, such as *omuhihi*, which are known to have other important utility values.

The recommendations are therefore as follows.

- The aim of field practitioners (extension workers, farmers, researchers, and other advocates) should be careful conservation of rare and valuable species and varieties, especially in their preferred niches (*in situ* conservation).
- Government at all levels should encourage and facilitate scientists to evaluate and document agrodiversity at regular intervals and to develop programmes for agrobiodiversity conservation.

- Local and central government should develop guidelines and regulations on balancing the marketing of food-crop produce with household food security.
- Extension workers and researchers should technically guide local communities on sustainable management and use of multi-purpose indigenous plants to avoid overexploitation.
- Researchers should examine possible contradictory government policies and plans (e.g. the liberalization of marketing of agricultural produce and the PMA), paying attention to the provisions of the CBD to which the Ugandan government is a signatory.

Theme two: Public awareness

Despite the many different values and functions of agrodiversity in general and agrobiodiversity in particular, most people are not aware of the need for conservation. No attention has been paid to agrodiversity in development plans and budgetary allocations, and consequently its status is threatened. Public awareness is essential in creating a commitment and positive attitude towards conservation and sustainable utilization of agrobiodiversity.

The following recommendations are therefore made.

- Relevant sector ministries, including the Ministry of Water, Lands, and Environment (MWLE) and the Ministry of Agriculture, Animal Industry, and Fisheries (MAAIF), should facilitate agrodiversity experts to carry out sensitization campaigns aimed at government officials at all levels and at the general public. Mass media and farmer-led public exhibitions are particularly effective in creating public awareness.
- Local governments should facilitate field practitioners and PLEC farmers to demonstrate good practices that promote agrobiodiversity conservation.

As an example of the effectiveness of practical demonstrations, these are some lessons learnt by different stakeholders from a farmer-led exhibition in Bushwere:

- methods of planting and preserving traditional plants
- relationships between land degradation and ethnicity
- planting trees with water containers for drip irrigation
- uses of various medicinal plants
- plants for feeding bees and making beehives
- how to conserve trees
- the importance of farmers and scientists working together
- importance of culture in agriculture and environment management and the need to maintain it
- how to store farm produce and seeds properly.

Theme three: Indigenous knowledge

The important issues are as follows.

- Rural communities are custodians of agrodiversity and the asso-ciated indigenous knowledge is of national and global significance. In Uganda, the diverse ethnicity and liberal intermarriages promote the diversity of such knowledge. Elderly people and women are the major sources of indigenous knowledge regarding agrodiversity. They are often skilled in the identification and utility values of different species and varieties, field management practices, and selection and preserva-tion of seed materials. However, very little of this indigenous knowl-edge is documented.
- The current social, cultural, and economic transformations in commu-nities, which increase generation gaps, pose a threat to the protection of such important knowledge. For instance, formal education keeps children away from home most of the time and the youths tend to sep-arate themselves from their parents at an early age (16 to 20 years). Also, indigenous knowledge on non-commercial agrobiodiversity is overlooked and thus not adopted by the young. These issues limit the dissemination of useful indigenous knowledge that would promote conservation and sustainable use of agrodiversity.
- Those with specialised indigenous knowledge, such as medicinal her-balists and artisans, are reluctant to share their knowledge for fear of piracy and exploitation.

The extent and importance of indigenous knowledge can be judged by the example of indigenous knowledge on seed preservation in Bushwere demonstration site.

- Hanging selected maize cobs (with or without sheath) and sorghum heads under the veranda or suspending them above fireplaces in the kitchen.
- Spreading Irish potato "seed" on the floor in a dark room.
- Partial roasting of sorghum grains using dry banana leaves and storing the grains in the resulting ash.
- Bagging beans and other grains and keeping the bags raised off the floor in residential houses or lockable stores.
- Mixing bean seeds with ash from burnt dry banana leaves or coating them with concentrated banana juice.
- Using cow *ghee* as a pest repellent for beans in store.
- Heaping well-dried unthreshed harvests of cereals like millet, sorghum, and maize in special traditional granaries called *ebihumi* or *ebitara*.

The recommendations are therefore as follows.

- All efforts to integrate environmental education into formal and non-formal education systems should emphasize indigenous knowledge, cultural values, and good land husbandry practices that promote agro-diversity conservation.

- Indigenous knowledge of biodiversity and land-use management should be promoted through participatory research, and documented by relevant stakeholders.
- All stakeholders should attach appropriate and effective incentives (material, monetary, or moral) to indigenous knowledge in order to enhance its dissemination.
- Government should put legal safeguards on indigenous knowledge by instituting and enforcing property and patent rights regulations.

Theme four: Capacity building

The main issues are as follows.
- Under the decentralization policy and plan for the modernization of agriculture, the government has increased access to field extension services at grassroots levels. Nevertheless, there is inadequate human capacity to integrate the concepts of agrodiversity conservation and environment protection into agricultural production.
- Farmers generally lack the ability to seek knowledge and advisory services at household and community levels.
  The recommendations are therefore as follows.
- Providers of advisory services should strengthen their capacity through training in the integration of agrodiversity conservation and environmental protection concepts into agricultural production and land management.
- Local governments should facilitate farmers and farmer groups skilled in practices that promote agrodiversity conservation to demonstrate to other farmers.
- The government – through the NAADS (National Agricultural Advisory Services) programme and other relevant agencies – should facilitate local communities to monitor and evaluate agrobiodiversity and enable them to demand advisory services by funding capacity building.
- When an appropriate agrodiversity conservation technique is identified, it should be studied and developed through regular interaction of and participatory evaluation by expert innovative farmers and scientists. This process promotes sharing of knowledge, effective learning experiences for both parties, and confidence of the expert farmers in their ability to demonstrate the good practices and train other farmers in developed skills.
- Initially, agro-technologies that yield value to the farmer while also conserving agrobiodiversity should be starting points for demonstrations. Both the farmers and the scientists therefore require substantial

funding, which the government (through the NAADS programme) should provide.
- Advisory service providers need to be trained in skills for integrating agrodiversity conservation into their extension work. This would enable them to guide farmers, for instance, on how to develop sustainable ways of integrating livestock, bees, and associated fodder species into crop production systems with special emphasis on complementarity of enterprises. The same people would enable the rural poor to add value to underutilized or threatened agrobiodiversity.

Theme five: Gender aspects

The main issues are as follows.
- Women in Uganda, by virtue of their production and reproduction roles and responsibilities, are the main custodians of agrodiversity, including agricultural biodiversity resources. However, gender imbalances at household level constrain efforts towards agrobiodiversity conservation. For instance, except in female-headed households, men control most production resources, especially land and capital, and also cash benefits accrued from sales of agrobiodiversity products. Women and children are left with crops and livestock that have little or no cash benefits, and if the benefits of such crops and livestock later increase, men commonly regain control. This limits the capacity of women and children to invest in long-term conservation strategies.
- The roles and responsibilities of rural men and women have made them, in effect, unwitting agents of agrobiodiversity erosion. The domestic role of women and their dependency on wood for fuel drives their role in deforestation. They may even use grass and crop residues for cooking. Similarly, over-dependency on the land for basic household necessities (food, shelter, and cash) and the associated roles and responsibilities of men lead them to erode agrobiodiversity through tree felling, bush burning, and stone, murram, clay, and sand mining.
The following recommendations are therefore made.
- Women should have increased access to and control of land, capital, and benefits as well as improved decision-making skills as an incentive to invest in agrobiodiversity conservation.
- The Ministry of Gender, Labour, and Social Development should facilitate gender sensitization of communities and local leaders while also providing technical guidance on demonstrating gender balance in agrobiodiversity conservation and utilization.
- Local leaders and field practitioners should ensure equitable participation of both men and women in agrodiversity conservation programmes in order to integrate the contributions of both.

- Men and women should be trained in technologies that minimize wood requirements for fuel and construction and in the management and use of multi-purpose trees and shrubs.

In other words, efforts towards gender balance and the empowerment of women as a vehicle for the enhancement of agrodiversity conservation should involve:

- promotion of husband and wife cooperation (joint planning)
- educating households (men, women, and children) in the requirements for promoting agrodiversity conservation resources, particularly in matters related to land use
- sensitization and training in integrating gender issues in all household activities to enhance equitable sharing of benefits from agrodiversity.

Theme six: Land aspects

The main issues are as follows.

- Despite the existence of an an environment management policy (1994) and regulations on the management of hilly and mountainous areas (Uganda Government 2000), neither of these mention preservation or sustainability of the rich diversity of agricultural and biological resources of these fragile agro-ecosystems. The regulations focus on conserving the soil and water resources. Even the proposed National Plan for Biodiversity Conservation, while catering for the different kinds of biodiversity, omits agricultural biodiversity.
- There is also a serious omission in the National Environment Statute (1995). When the government needs to take over privately owned land, compensation for biodiversity other than conventional crops is overlooked.
- Another issue is the absence of national policies on soils and land use, which creates a policy vacuum on land use and management activities. A case in point is the widespread mining of clay, murram, sand, soil, and stones, which goes unrestricted.

The following recommendations are therefore made.

- Local governments should formulate by-laws to support national regulations covering the management of hilly and mountainous areas and any other relevant policies and guidelines available.
- The central government should amend the Environment Statute of 1995 to include other biodiversity conserved on land besides crops in its compensation package, as well as providing guidelines for the effective valuation of such resources.
- The central government should accelerate the formulation of a national land-use policy, a national soils policy, and a biodiversity conservation action plan, which should all include the conservation of agrodiversity.

- The forthcoming land-use policy should deliberately include the rehabilitation of mined and degraded lands, using appropriate species.

Theme seven: Institutional aspects

The main issues are as follows.
- Although the Ugandan government (through the NEMA) and GEF (through the UNDP, UNEP, and the World Bank) has facilitated work on biodiversity by providing a structure for its monitoring and evaluation, or by funding short courses and projects, there is still little focus on agricultural biodiversity. Emphasis is currently on biodiversity in protected areas and natural ecosystems, but not on agricultural lands. Consequently, data collection and processing (and hence data custody on agrobiodiversity in particular and agrodiversity in general) are lacking.
- Agrodiversity is multisectoral, as reflected in the diversity of issues presented in the preceding themes. Consequently its management cannot be handled solely by one of the existing ministries/agencies. This calls for strategic coordination and networking of institutions and individuals that do work related to agrodiversity.
- There are no active grassroots environmental organizations to implement the existing relevant policies and recommendations.
  The following recommendations are therefore made.
- The Faculty of Agriculture at Makerere University should continue its work on compiling a database and monitoring and evaluation, as well as technology development and providing technical guidance to the government for appropriate policy development.
- The NEMA (in collaboration with higher education institutions) should coordinate networking through seminars for all institutions and agencies working on biodiversity, including agrobiodiversity.
- Local governments, NGOs, and community-based organizations should prioritize agrobiodiversity conservation programmes, including monitoring and evaluation.
- Central government and the NEMA in particular should speed up the development of regulations and guidelines on the rights of access to and patents of biological and genetic resources. These should emphasize agrodiversity and should be disseminated to all stakeholders.
- Districts should set up technical committees (involving relevant stakeholders) coordinated by the district environment officers (DEOs) to oversee agrodiversity issues and ensure that conservation is embedded in their work plans and budget. These committees should also take the lead in implementing existing policies, laws, and regulations with sections relevant to biodiversity and agrodiversity conservation.

Theme eight: Technology development, transfer, and adoption

The most important issues are as follows.

- The concept of agrodiversity is new in Uganda. It is multi-dimensional in nature and few related technologies have yet been developed. However, projects such as PLEC have initiated capacity-building efforts using demonstration sites. These aim to enable farmers, as guardians of agrodiversity, to play a central role in appreciating its benefits and then developing and disseminating appropriate technologies for its conservation and sustainable use. This was done through collaboration with innovative farmers who agreed to carry out on-farm experimentation and demonstrate good practices. Subsequently, farmers were encouraged to form common-interest groups for agrodiversity conservation. Emphasis on equitable participation and the sharing of benefits between gender and common-interest groups was encouraged as a means of ensuring sustainability, transfer, and adoption of the developed technologies. The project has involved local leaders, community members, policy/decision-makers, and other stakeholders in evaluating the technologies at various levels of their development. Several problems, however, have been noted and need to be addressed. There is a lack of expertise in the economic valuation of agrodiversity to highlight the financial benefits/returns of its conservation. This hinders technology transfer and adoption. Further, the education system keeps children off-farm most of the time. Coupled with the trend for youths to distance themselves from their homes, this has weakened farm apprenticeship and affected technology transfer to and adoption by the young.
- Over-dependency on rainfall, low groundwater tables, and high desiccating winds of the hilly and mountainous areas limit adoption and sustainability of most agrobiodiversity conservation technologies. This is because plants on hillsides and hilltops suffer serious water stress during the dry seasons. This is particularly severe where the water-holding capacity of the soils is low, as in the Mbarara highlands.
- The PLEC project work carried out at the Bushwere demonstration site requires continuity. It should also be extended to other farmers and parts of Uganda, particularly hilly and mountainous areas. The work of PLEC-Uganda is acknowledged by different stakeholders. Professor Edwin A. Gyasi concluded after visiting Uganda that "EA-PLEC sits on a mine of agrodiversity in East Africa. PLEC-Uganda has started laying foundations for development of methodologies towards conservation of the diversity."

Mr Bigirimana, DAO representative, said as a guest speaker at a stakeholders' workshop:

I have been impressed by farmers' demonstrations and experiments and I think some deserve to be called doctors and professors. In the early days, researchers used to work at research stations, i.e. Kawanda, Namulonge, Serere, Kalengyere, and Makerere University Farm at Kabanyolo, only and never used to reach farmers. If there were new hybrid varieties released, they would end up at the district agricultural officers' offices and never adequately reach the people at the grass roots. I am happy that Makerere University staff, through the PLEC project, have demonstrated that this tendency can be changed.

Nkwasibwe Denis, Chairman LC III, Mwizi subcounty, as a local host of the stakeholders' workshop, said: "I have been impressed by the improvements that I have seen on the farmers' gardens and subsequently their welfare in homes. I thank them for their active participation in biodiversity conservation." Bushwere Women's Group, in their song at a farmer-led exhibition of agrodiversity potential (August 2000) said: "We, the farmers of Bushwere, have learnt a lot from collaborating with PLEC scientists. We have learnt how to manage many crop types in our gardens, to restore and preserve our traditional crops and varieties, how to make money and protect our environment using different plants grown on our land." (Direct translation from Rukiiga language.)

The recommendations are therefore as follows.

- The government should retrain existing providers of extension services and facilitate higher education in agrodiversity technology development, transfer, and adoption as well as development of techniques for profiling and valuing biodiversity and its conservation.
- Researchers should use a multidisciplinary approach involving farmers, socio-economists, policy-makers, and implementers when developing agrodiversity technologies.
- The government and the private sector should support local communities in their quest for running water (for home and agricultural use). This could be via windmills, which pump water from large reservoirs and use gravitational flow to distribute water to homes and gardens.
- The government, private agencies, and donor communities should fund the continuation of PLEC initiatives in Bushwere and the extension of its agrodiversity conservation work to other parts of the country.
- The present process of integrating environmental education into the school curriculum by the NEMA and the Ministry of Education and Sports should emphasize agrodiversity conservation.
- Similarly, the environmental education strategy being formulated by the NEMA should emphasize good land husbandry practice that promotes agrobiodiversity conservation.

- The Faculty of Agriculture at Makerere University should introduce a course unit on agrodiversity.
- For immediate transfer and adoption, PLEC should present findings to the district council and hold a field-day/exhibition for all field extension staff in the district.

Conclusion

Considering the many issues raised and recommendations made on agrodiversity conservation, it is imperative that Uganda's government and other stakeholders should facilitate the development of a national agrodiversity policy to promote sustainable management of the important agricultural and biological resources.

In summary, the major lessons learnt and limitations to policy formulation are as follows.

- There is appreciation of PLEC approaches to agricultural resource management and efforts towards developing policy and technical recommendations for agrobiodiversity conservation. Government officials, the national GEF programme, NEMA, farmers, NGOs, education institutes, and other stakeholders have all clearly expressed the value of approaches pioneered by PLEC.
- Agrobiodiversity is a multisectoral resource, which makes it more difficult to identify the appropriate institution, authority, or body to take responsibility for the developed policy and technical recommendations.
- It is difficult to convince people – especially policy-makers – that PLEC's work with a few farmers in one parish is a sufficient foundation on which to base policy. They may feel that the project should be extended before conclusions are drawn.

REFERENCE

Uganda Government. 2000. *The Uganda Gazette*, Vol. XCIII, No. 5, Entebbe: UPPC.

Appendix

East Africa PLEC General Meeting, 26–28 November 2001

List of participants

Name	Country
Charles Nkwiine	Uganda
Joy Tumuhairwe	Uganda
Jovia Manzi	Uganda
Frank K. Muhwezi	Uganda
Fred Tuhimbisibwe	Uganda
John Kang'ara	Kenya
Ezekiah Ngoroi	Kenya
Kajuju Kaburu	Kenya
Charles Rimui	Kenya
Bernard Njeru Reuben Njiru	Kenya
Michael Stocking	United Kingdom
Fidelis Kaihura	Tanzania
Jerry Ngailo	Tanzania
Barnabas Kiwambo	Tanzania
John Elia	Tanzania
Peter Kapingu	Tanzania
Essau Mwalukasa	Tanzania
Edward Ngatunga	Tanzania

Non-resident participants

Gidiel Loivoi	Arumeru – Ng'iresi
Frida Kipuyo	Arumeru – Kiserian
Deusdedit Rugangira	Arumeru
Beatrice Maganga	Mwanza
Cypridion Maganga	Mwanza
Edina Kahembe	Arumeru

Opening speech by the guest of honour, Wilfred Ngirwa, Permanent Secretary, Ministry of Agriculture and Food Security

Mr Chairman,
The scientific coordinator for PLEC,
East Africa PLEC scientists and farmers,
Honourable guests,
Ladies and gentlemen,

Mr Chairman, Tanzania is privileged to host the East Africa PLEC General Meeting after four years of its existence as a project. I also feel privileged to be with you today for the opening occasion.

On behalf of the Government of Tanzania, the Ministry of Agriculture and Food Security, and the Arumeru farmers, I have the honour to welcome you to Tanzania and Arusha in particular.

Mr Chairman, this is a remarkable and important gathering of East African farmers, scientists, and policy-makers from Tanzania, Uganda, and Kenya to deliberate on PLEC, which is a project on People, Land Management, and Environmental Change in rural farming communities.

Your meeting will let you share experiences and be able to know if PLEC has achieved its objectives in areas of:

- establishment of historical and baseline comparative information on biodiversity and agrodiversity at the landscape level in representative diverse and dynamic sites
- development of participatory and sustainable models of biodiversity management based on farmers' technologies and knowledge within agricultural systems at the community and landscape levels
- development of recommendations on approaches and policies for sustainable agrodiversity management to key government decision-makers, farmers, and field practitioners
- capacity building, particularly in the area of agrodiversity.

Mr Chairman, I request you to listen carefully to the farmers attending this meeting who are the target for this project (PLEC), who have been fully involved in development of participatory and sustainable models of biodiversity management in their own fields and at community level. They will be able to tell you about the success, the shortcomings, and the corrective measures for this project.

It is now an unquestionable fact that farmers, researchers, and extension workers have something to learn from each other. This is evident within PLEC whereby expert farmers have emerged and acted as teachers (extension agents) of other farmers and data providers to scientists. This interaction has successfully enabled achievement towards successful technology development and dissemination.

The work done at the PLEC sites in East Africa has resulted in the formulation of recommendations that aim to sustain crop and management diversity and improved local livelihoods. This is a new approach, which needs to be replicated in other new areas in the agriculture sector programmes in our countries.

Mr Chairman, as you know, Kenya, Uganda, and Tanzania are basically agricultural-dependent economies dominated by smallholder farmers who live in rural areas. Poverty studies have shown that most of the poor people live in the rural areas (and these are the smallholder farmers). Therefore, national poverty alleviation strategies need to be focused to address the problems rural farmers are facing.

Mr Chairman, the Ministry of Agriculture and Food Security considers PLEC as an important model and contributor in enhancement of food security and alleviation of rural poverty.

Consequently, I would like to thank the United Nations University for their role in the implementation of this project, the Global Environmental Facility for funding the project, the project management for including East Africa among project implementing regions, and ARI Ukiriguru for coordinating PLEC work in Tanzania. Special thanks to the project management for including farmers in such meetings where they really present their first-hand experience and ideas and defend their interests.

I would also like to thank PLEC hosts in Arumeru, mainly the District Executive Director, the Zonal Director Research and Development, Northern Zone, and the District Agricultural and Livestock Development Officer, Arumeru for facilitating the work; also the researchers, extension workers, expert farmers, and other personnel who devoted their time and skills to ensuring the success of the project; and finally, the hardworking farmers in PLEC sites for their dedication.

Mr Chairman, before I conclude, may I remind you that at the end of this meeting we expect you to come up with clear and pragmatic recommendations on the way forward. Once again, I warmly welcome you. I wish you a successful meeting and I hope you will find Arusha as a home place for your follow-up meetings.

Mr Chairman, I now have the honour and indeed pleasure to declare the PLEC General Meeting officially open. Thank you for your attention.

Closing statement by Robert Kileo (ZRC-L)

Mr Chairman,
PLEC scientific coordinator,
Representative, Ministry of Agriculture and Food Security,
Meeting organizers,
Implementing scientific staff and farmers,
Invited guests,
Ladies and gentlemen,

First of all, on behalf of the Zonal Director, Research and Development in the Lake Zone, and myself, I thank the organizers of this important meeting for the opportunity to attend and deliver a closing statement. I find myself honoured.

We have heard a series of field experience presentations of various PLEC activities. The most exciting thing to me was the involvement of mixed stakeholders. We received presentations from leading scientists (PLEC) and collaborating scientists as well as farmers. This is one of the many indications that PLEC is participatory in practice. I commend you all for this and encourage you to continue with that spirit. It should also be noted that findings from all three East African countries (Kenya, Tanzania, and Uganda) were presented and discussed.

Dear organizers, it is not my aim in this closing statement to go through what was presented in the past three days. However, it was gratifying to hear of a range of activities presented. There was a thorough coverage of PLEC methodologies, botanical knowledge and its utilization, and detailed case studies in all participating countries as well as socio-economic aspects influencing farmers' decisions and utilization of biodiversity potentials. Other issues included policy and farmers' indigenous knowledge. To me this was again a clear indication that the PLEC methodology is a holistic approach that deals with most farmers' circumstances simultaneously. By doing so, it is likely that most farmers' problems can be solved within a short period.

Dear organizers, I am also obliged to comment on what we observed during the field visit. Several technologies were being demonstrated on-farm. We saw farmers working hard to overcome existing environmental challenges by practising a diversity of activities (crop and livestock). It was clear that farmers have a very wide range of uses of available plant species. This is a positive sign and again I congratulate you for that effort. All these efforts to me seem to be an indication that PLEC activities have contributed to farmers' income, food security, and environmental conservation. I hope that the few farmers who are now participating will be the catalysts and eventually these technologies will be disseminated to other farmers and places.

Together with these successes, PLEC still faces some challenges. We have heard from case studies that farming communities in the Eastern African highlands face some threats. These include increase in the human population, land degradation, and harsh weather conditions. PLEC should work harder on these by developing suitable technologies to address these challenges.

Dear organizers, I am informed that you are going to have discussions on proposal development for PLEC's future. This is an important step. I request you to incorporate deliberations raised during discussions. I believe that it is possible to include some (for example, scaling up, diversified donor funds, and so on).

Lastly, I am delighted to hear that all expected/invited participants attended this meeting. I thank you all for that and encourage you to keep it up. I am aware that some of the participants came all the way from Uganda, Kenya, Arusha, Tanga, and Mwanza. After this meeting, I wish you a safe journey back home or to your working places. Thank you for participating into this meeting and listening to me.

After these few statements, and on behalf of the Zonal Director Research and Development, Lake Zone, I formally declare the meeting closed.

Again, thank you very much.

List of contributors

Kenya

John N. N. Kang'ara, Principal research officer with Kenya Agricultural Research Institute (KARI), head of livestock research section, and the PLEC team leader in Embu, Kenya, Kenya Agricultural Research Institute (KARI) RRC Embu, Box 27 Embu, Kenya.

Kajuju Kaburu, Agronomist/ nutritionist, Kenya Agricultural Research Institute (KARI) RRC Embu, Box 27 Embu, Kenya.

Charles M. Rimui, Senior technician in soil and water management, Kenya Agricultural Research Institute (KARI) RRC Embu, Box 27 Embu, Kenya.

Ezekiah H. Ngoroi, Senior research officer with KARI Embu, Kenya Agricultural Research Institute (KARI) RRC Embu, Box 27 Embu, Kenya.

Seth Amboga, Research officer/ statistician, Kenya Agricultural Research Institute (KARI) RRC Embu, Box 27 Embu, Kenya.

Kaburu M'Ribu, Professor of Horticulture at Kenya Methodist University (KEMU), Kenya Methodist University, PO Box 267, Meru, Kenya.

Barrack Okoba, Senior research officer with KARI Embu; soil and water management scientist, Kenya Agricultural Research Institute (KARI) RRC Embu, Box 27 Embu, Kenya.

Julius M. Muturi, Veterinary research officer, Kenya Agricultural Research Institute (KARI) RRC Embu, Box 27 Embu, Kenya.

Francis K. Ngugi, Senior technical officer, livestock/agroforestry, Kenya Agricultural Research Institute (KARI) RRC Embu, Box 27 Embu, Kenya.

Immaculate Mwangi, Senior technician in agronomy, Kenya Agricultural

Research Institute (KARI) RRC Embu, Box 27 Embu, Kenya.

Bernard N. Njiru, Retired primary school teacher and farmer at Nduuri Kenya PLEC site, c/o Nduuri Primary School, PO Runyenjes, Nduuri, Embu, Kenya.

Tanzania

Essau E. Mwalukasa, Agricultural research officer at the Agricultural Research and Development Institute, Ukiriguru, PO Box 1433, Mwanza, Tanzania.

Jerry A. Ngailo, Agricultural research officer at the Agricultural Research and Development Institute, Mlingano, PO Box 5088, Tanga, Tanzania.

Freddy P. Baijukya, Agricultural research officer at the Agricultural Research and Development Institute, Maruku, PO Box 127, Bukoba, Tanzania.

Barnabas J. Kiwambo, Senior agricultural research officer at the Agricultural Research and Development Institute, Mlingano, PO Box 5088, Tanga, Tanzania.

Robert M. L. Kingamkono, Department of Agricultural Engineering and Land Use

Planning, PO Box 3003, Morogoro, Tanzania.

Deusdedit M. Rugangira, District agricultural and livestock development officer, PO Box 2416, Arusha, Tanzania.

Edina Kahembe, Field officer, District Agriculture and Livestock Development Office, Arumeru, Arusha, PO Box 2416, Arusha, Tanzania.

Frida P. Kipuyo, Expert farmer, livestock diversity, Kiserian Village, Arumeru, Arusha, Tanzania.

Kisyoki Sambweti, Expert farmer, natural pasture conservation and management, Kiserian Chini Village, Arumeru, Arusha, Tanzania.

Gidiel L. Loivoi, Expert farmer, agricultural intensification and crop livestock integration, Ng'iresi Village, Arumeru, Arusha, Tanzania.

Uganda

Joy K. Tumuhairwe, PLEC researcher/ team leader and agriculturalist/ graduate teacher/soil scientist

(pedology and land use), Faculty of Agriculture, Makerere University, PO Box 7062, Kampala, Uganda.

Charles Nkwiine, PLEC researcher and agriculturalist/soil scientist (agrobiology), Faculty of Agriculture, Makerere University, PO Box 7062, Kampala, Uganda.

John B. Kawongolo, PLEC researcher and agricultural engineer (post-harvest), Faculty of Agriculture, Makerere University, PO Box 7062, Kampala, Uganda.

Francis K. Tumuhairwe, PLEC researcher and economist (agricultural economics), Faculty of Agriculture, Makerere University, PO Box 7062, Kampala, Uganda.

Chris Gumisiriza, Trainee research assistant for PLEC and agriculturalist, Faculty of Agriculture, Makerere University, PO Box 7062, Kampala, Uganda.

Jovia Nuwagaba-Manzi, Trainee research assistant for PLEC and agricultural extensionist, Faculty of Agriculture, Makerere University, PO Box 7062, Kampala, Uganda.

Frank Muhwezi Kashaija, Farmer (retired primary school teacher), Chikunda Primary School, Bushwere Parish, Mwizi Subcounty, Mbarara, Uganda.

Fred Tuhimbisibwe, Farmer (retired soldier), C/o Chikunda Primary School, Bushwere, Mwizi Subcounty, Mbarara, Uganda.

Editors

Fidelis B. S. Kaihura, Senior soil scientist and head of Natural Resources Management Programme at the Agricultural Research and Development Institute, Ukiriguru, Mwanza, Tanzania, PO Box 1433, Mwanza, Tanzania.

Michael A. Stocking, Professor of Natural Resources at the University of East Anglia, UK, associate scientific coordinator for PLEC, University of East Anglia Norwich, Norfolk, NR4 7TJ UK.

Index

240

Catalogue Request

Name: _____

Address: _____

Tel: _____

Fax: _____

E-mail: _____

To receive a catalogue of UNU Press publications kindly photocopy this form and send or fax it back to us with your details. You can also e-mail us this information. Please put "Mailing List" in the subject line.

**United Nations
University Press**

53-70, Jingumae 5-chome
Shibuya-ku, Tokyo 150-8925, Japan
Tel: +81-3-3499-2811 Fax: +81-3-3406-7345
E-mail: sales@hq.unu.edu http://www.unu.edu